Simple Methods to Study Pedology and Edaphology of Indian Tropical Soils

D. K. Pal

Simple Methods to Study Pedology and Edaphology of Indian Tropical Soils

D. K. Pal
ICAR-NBSS&LUP
Nagpur, Maharashtra, India

ISBN 978-3-319-89598-7   ISBN 978-3-319-89599-4 (eBook)
https://doi.org/10.1007/978-3-319-89599-4

Library of Congress Control Number: 2018940437

© Springer International Publishing AG, part of Springer Nature 2019
This work is subject to copyright. All rights are reserved by the Publisher, whether the whole or part of the material is concerned, specifically the rights of translation, reprinting, reuse of illustrations, recitation, broadcasting, reproduction on microfilms or in any other physical way, and transmission or information storage and retrieval, electronic adaptation, computer software, or by similar or dissimilar methodology now known or hereafter developed.
The use of general descriptive names, registered names, trademarks, service marks, etc. in this publication does not imply, even in the absence of a specific statement, that such names are exempt from the relevant protective laws and regulations and therefore free for general use.
The publisher, the authors and the editors are safe to assume that the advice and information in this book are believed to be true and accurate at the date of publication. Neither the publisher nor the authors or the editors give a warranty, express or implied, with respect to the material contained herein or for any errors or omissions that may have been made. The publisher remains neutral with regard to jurisdictional claims in published maps and institutional affiliations.

Printed on acid-free paper

This Springer imprint is published by the registered company Springer International Publishing AG part of Springer Nature.
The registered company address is: Gewerbestrasse 11, 6330 Cham, Switzerland

I dedicate this book to my parents and parents-in-law

Preface

Although much valuable work is done throughout the tropics, difficulties still remain to manage these soils to sustain their productivity due to the long absence of comprehensive knowledge on their formation. In any nation, soil care needs to be a constant research endeavor to ensure agricultural productivity and self-sufficiency including Indian subcontinent. New soil knowledge base becomes critical when attempts are made to fill the gap between food production and future population growth. Realizing this urgency, research efforts during the last few decades were undertaken on the benchmark and identified soil series by the Indian pedologists and earth scientists. Their work has been commendable and they have managed to establish an organic link between pedology (soils of the past and the present), mineralogy, taxonomy, and edaphology of five pedogenetically important soil orders like Alfisols, Mollisols, Ultisols, Vertisols, and Inceptisols of tropical Indian environments. This knowledge was enriched when the focus of soil research changed qualitatively due to the use of high-resolution mineralogical, micromorphological, and age-control tools along with the geomorphic and climatic history of soil formation in the present and past geological periods. This advancement in basic and fundamental knowledge on Indian tropical soils provided unique perspectives to develop several index soil properties as simple methods to study their pedology and edaphology. Continued efforts in basic pedological research are needed to understand some of the unresolved edaphological aspects of Indian tropical soils especially when soil properties are modified due to climate change and some layer silicate minerals formed in the present and past climates are preserved. Application of many of such simple methods would be of tremendous significance even in the absence of high-end instrumental facilities, to study pedology and mineralogy. More than one-third of the soils of the world are tropical soils, and thus the recent advances in developing simple and ingenious methods to study pedology and edaphology of Indian tropical soils (Entisols, Inceptisols, Mollisols, Alfisols, Vertisols, and Ultisols) may also be adopted by both graduate students and young soil researchers for similar soils elsewhere for an expeditious development of national soil information system to enhance crop productivity and maintain soil health in the twenty-first century.

In each of the ten chapters here, a specific theme has been chosen to showcase the worthiness of applying such simple methods that have emerged from rigorous research done mainly at the ICAR-NBSS&LUP, Nagpur, India. The author would like to acknowledge the valuable contributions of mentors and peers who have helped along the tenuous path of research. He will remain ever grateful to late Prof. S.K. Mukherjee and late Prof. B. B. Roy, who held the coveted position of Acharya P.C. Ray Professor of Agricultural Chemistry at the University of Calcutta, for drawing him to soil research that offered more than a lifetime of fascinating problems to unravel. The emergence of simple and ingenious methods to study pedology and edaphology as depicted in the book has been possible due to the significant research contributions made through a joint endeavor made by the author along with his esteemed colleagues. They are Drs. T. Bhattacharyya, P. Chandran, S. K. Ray, and Pramod Tiwari of ICAR-NBSS&LUP, Nagpur, and Prof. Pankaj Srivastava of Geology Department, Delhi University, and also several M.Sc. and Ph.D. students at ICAR-NBSS&LUP. Unstinted technical support and assistance received from Messrs. S.L. Durge, G.K. Kamble, and L.M. Kharbikar helped the author enormously in bringing the task to successful fruition.

The author duly acknowledges the sources of the diagrams and tables that have been adapted mostly from his publications.

The author is grateful to his wife, Banani, his daughters Deedhiti and Deepanwita, his brother-in-law Dhrubajyoti, and his sons-in-law Jai and Nachiket, for their patience, understanding, encouragement, and above all unwavering moral support.

Nagpur, India   D. K. Pal

Contents

1 **Methods to Study Pedology and Edaphology of Indian Tropical Soils: An Overview** 1
  References 4

2 **Evidence of Clay Illuviation in Soils of the Indo-Gangetic Alluvial Plains (IGP) and Red Ferruginous (RF) Soils** 7
  2.1 Introduction 8
  2.2 Depth Function of Clay Mica: An Incontrovertible Evidence of Clay Illuviation in IGP Soils 9
    2.2.1 Development of a Simple Analytical Method 9
  2.3 Clay Illuviation in Red Ferruginous (RF) Soils of the Humid Tropical (HT) vis-a-vis IGP Soils of Semi-Arid (SAT) Climate 10
  2.4 Micro-Morphological Thin Section Studies to Resolve the Argillic Horizon Issue 11
    2.4.1 Simple Analytical Way to Recognize the Argillic horizon in Indian Tropical Soils 14
  References 16

3 **Clay Illuviation and Pedoturbation in SAT Vertisols** 19
  3.1 Introduction 19
  3.2 Uniqueness of Clay Illuviation in Highly Smectitic Vertisols 21
  3.3 Simple Tool to Establish Clay Illuviation as the Primary and Pedoturbation as Partially Functional Pedogenic Process in Vertisols 23
  References 24

4 **Cracking Depths in Indian Vertisols: Evidence of Holocene Climate Change** 27
  References 31

5 **Unique Depth Distribution of Clays in SAT Alfisols: Evidence of Landscape Modifications** 33
  References 37

6	**Easy Identifications of Soil Modifiers**	39
	6.1 Introduction	39
	6.2 Gypsum and Its Significance in Redefining Saline-Sodic Soils and Managing SAT Vertisols	40
	6.3 Pedogenic Calcium Carbonate (PC) and Its Role in Land Evaluation	41
	6.4 Zeolites and their Role in Soil Formation and Management	42
	6.4.1 Analcime in IGP Soils	42
	6.4.2 Heulandite in Soils of the Deccan Basalt Areas	43
	6.4.3 Simple Analytical Method to Identify Zeolites in Soils	43
	6.5 Palygorskite and its Role in Soil Classification and Management	44
	6.5.1 Simple Method to Identify the Presence of Palygorskite	46
	References	46
7	**Mineralogy Class of Indian Tropical Soils**	49
	References	54
8	**Hydraulic Conductivity to Evaluate the SAT Vertisols for Deep-Rooted Crops**	57
	References	62
9	**Clay and Other Minerals in Selected Edaphological Issues**	63
	9.1 Introduction	63
	9.2 Carbon Sequestration in Indian Tropical Soils	64
	9.3 Acidity, Al Toxicity, Lime Requirement and Phosphate Fixation in Soils of Indian Tropical Environments	68
	9.4 K Release from Biotite Mica and Adsorption of Nitrogen and Potassium by Vermiculite of Indian Tropical Soils	70
	References	71
10	**A Critique on Degradation of HT and SAT Soils in View of Their Pedology and Mineralogy**	75
	References	79
11	**Anomalous Potassium Release and Adsorption Reactions: Evidence of Polygenesis of Tropical Indian Soils**	81
	11.1 Introduction	82
	11.2 K Release and Adsorption in Polygenetic IGP Soils	82
	11.3 K Release and Adsorption in Polygenetic Vertisols	85
	11.4 K Release and Adsorption in Polygenetic RF Soils	88
	References	91
12	**Concluding Remarks**	93
	Index	95

About the Author

D. K. Pal graduated in 1968 with honours in Chemistry, obtained M. Sc (Ag) degree in Agricultural Chemistry with specialization in Soil Science in 1970 and earned his Ph. D. degree in Agricultural Chemistry in 1976 from the Calcutta University. He worked as a DAAD Post-Doctoral Fellow at the Institute of Soil Science, University of Hannover, West Germany during 1980–81.

Research activities of Dr. Pal have spanned more than three and a half decades and have focused on the alluvial (Indo-Gangetic Alluvial Plains, IGP), red ferruginous and shrink-swell soils of the tropical environments of India. His work has expanded the basic knowledge in pedology, paleopedology, soil taxonomy, soil mineralogy, soil micromorphology and edaphology. He also created an internationally recognized school of thought on the development and management of the Indian tropical soils as evidenced by significant publications in several leading international journals in soil, clay and earth sciences inviting critical appreciation by his peers.

He has built an excellent research team in mineralogy, micromorphology, pedology and paleopedology in the country. Under his guidance, the team members have carved out a niche for themselves in soil research at national and international level. He has mentored several M. Sc and Ph. D students of Land Resource Management (LRM) of the Indian Council of Agricultural Research- National Bureau of Soil Survey and Land Use Planning (ICAR-NBSS&LUP) under the academic programme at Dr. Panjab Rao Deshmukh Krishi Vidyapeeth (Dr. P D K V), Akola.

He has delivered numerous prestigious invited lectures at national and international meets. In addition, he has served as a reviewer for many journals of national and international repute, and contributed reviews and book chapters for national and international publishers. Recently he has authored a book 'A Treatise of Indian and Tropical Soils' published in January, 2017 by Springer International Publishing AG, Switzerland. He has been co- editor of books, proceedings and journals. He is the life member of many professional national societies in soil and earth science. He has also been conferred with many awards (The Platinum Jubilee Commemoration Award of the Indian Society of Soil Science, New Delhi for the year 2012, ICAR Award, Outstanding Interdisciplinary Team Research in Agriculture and Allied

Sciences, Biennium 2005–2006, and the 12th International Congress Commemoration Award, Indian Society of Soil Science, 1997) and fellowships (Honorary Member, The Clay Minerals Society of India, New Delhi, 2016, West Bengal Academy of Science and Technology, 2014, National Academy of Agricultural Sciences, New Delhi, 2010, Indian Society of Soil Science, New Delhi, 2001 and Maharashtra Academy of Sciences, Pune, 1996).

He worked as the Principal Scientist and Head, Division of Soil Resource Studies, ICAR-NBSS & LUP, Nagpur and as a visiting scientist at the International Crops Research Institute for the Semi-Arid Tropics (ICRISAT), Telangana.

Chapter 1
Methods to Study Pedology and Edaphology of Indian Tropical Soils: An Overview

Abstract The major tropical soils of India are Vertisols, Mollisols, Alfisols, Ultisols, Aridisols, Inceptisols and Entisols. Thus, India has the large variety of soils. These soils are not confined to a single production system and contribute enough to self-sufficiency in food production and food stocks of the Indian subcontinent. This could be achieved because of favourable natural endowment of soils, which remained highly responsive for the last several decades to management interventions advocated through National Agricultural Research Systems (NARS). In the initial years' of soil research in India, much valuable work was done but the knowledge base was not enough to manage these soils to sustain their productivity. Soil knowledge base becomes critical when attempts are made to fill the gap between food production and future population growth. This demands that soil care needs to be a constant research agenda in the Indian context. In responding to such national need, research endeavour during the last few decades on the benchmark soils by the Indian pedologists and earth scientists has been commendable when the focus of soil research changed qualitatively due to the use of high resolution mineralogical, micromorphological and age-control tools along with the geomorphic and climatic history. This change in focus resulted in acquiring new knowledge on basic and fundamental aspects on major soil types of the country, which provided unique guiding principles to develop several simple diagnostic analytical methods to identify the important pedogenic processes, mineral transformation, taxonomic rationale and solving some of their queer edaphological issues. The worthiness of use of simple methods are demonstrated in the following few chapters. It is hoped that the adaptation of the simple methods will facilitate an early completion of the long standing national robust national soil information system.

Keywords Indian tropical soils · Study of pedology and edaphology · Simple methods

Major part of the land area in India is in the region lying between the Tropic of Cancer and Tropic of Capricorn, and the soils therein are termed "tropical soils". Many, however, think of tropical soils as the soils of the hot and humid tropics only, exemplified by deep red and highly weathered soils, which are often thought improperly to be either agriculturally poor or virtually useless (Sanchez 1976; Eswaran et al. 1992). India has 5 distinct bioclimatic systems (Bhattacharjee et al. 1982) with varying mean annual rainfall (MAR), and the major soils of India are Vertisols, Mollisols, Alfisols, Ultisols, Aridisols, Inceptisols and Entisols covering 8.1%, 0.5%, 12.8%, 2.6%, 4.1%, 39.4% and 23.9%, respectively of the total geographical area (TGA) of the country (Bhattacharyya et al. 2009). Vertisols belong to arid hot, semi-arid, sub-humid and humid to per-humid climatic environments (Bhattacharyya et al. 2005; Pal et al. 2009). Mollisols belong to sub-humid and also humid to per-humid climates (Bhattacharyya et al. 2006). Alfisols belong to semi-arid, sub-humid and also in humid to per-humid climates (Pal et al. 1989, 1994, 2003; Bhattacharyya et al. 1993, 1999), whereas Ultisols belong to only humid to per-humid climates (Bhattacharyya et al. 2000; Chandran et al. 2005). Both Entisols and Inceptisols belong to all the 5 categories of bio-climatic zones of India, and Aridisols belong mainly to arid climatic environments (Bhattacharyya et al. 2008). This baseline information indicates that except for the Ultisols and Aridisols, the rest 5 soil orders exist in more than one bio-climatic zones of India, and the absence of Oxisols, suggests that soil diversity in India is large. Therefore, India can be called a land of paradoxes because of the large variety of soils and any generalizations about tropical soils are unlikely to have wider applicability in the Indian subcontinent (Pal et al. 2012, 2014, 2015; Srivastava et al. 2015). These soils are not confined to a single production system and contribute substantially to India's growing self-sufficiency in food production and food stocks (Pal et al. 2015; Bhattacharyya et al. 2014). This has been made possible because of favourable natural endowment of soils that remained highly responsive to management interventions made mainly by farming communities with the support from national and international institutions. India primarily being an agrarian society, soil care through recommended practices has been the engine of economic development, eliminating large portion of poverty and considerable transformation of rural communities (Pal et al. 2012, 2014, 2015; Srivastava et al. 2015).

In the past, much valuable work has been done throughout the tropics, but it has been always difficult to manage these soils to sustain their productivity and it is more so when comprehensive knowledge on their formation remained incomplete for a long time. Soil care continues to be the main issue in national development and thus needs to be a constant research agenda in the Indian context. This is imperative since soil knowledge base becomes critical when attempts are made to fill the gap between food production and future population growth. In this task basic pedological research is required to understand some of the unresolved edaphological aspects of the tropical Indian soils to develop improved management practices. In India, students of pedology (soil science) and pedo-geomorphology generally come across extreme difficulties in relating to examples applying the principles of soil science from text books devoted almost exclusively to soils of temperate climate of erstwhile Soviet Union, Europe and the USA. Realizing this predicament, research endeavour during the last few decades, undertaken on the benchmark and identified soil series by the Indian pedologists and earth scientists of well-known institutions, especially by those in the Indian Council of Agricultural Research-National Bureau of Soil Survey and Land Use Planning (ICAR-NBSS&LUP), has been commendable (Pal 2017). Scientists improvised existing analytical methods that are used to study pedology in different parts of temperate humid climate. These analytical methods were made compatible to Indian tropical soils while ascertaining pedogenic processes like clay illuviation, pedoturbation, and chemical degradation. Through significant research efforts, scientists have demonstrated that such improvised methods have become effective in explaining the basic pedogenetic processes mainly of the SAT soils (Pal 2017).

Scientists have used Vertisols as sign of Holocene climate change, pedogenic carbonates as index property of soil chemical degradation. Also they vigorously researched to highlight the clay minerals as they exist in soil environment, in relating many pedological and edaphological properties and identified polygenesis in adsorption and desorption behaviour of nutrient element. Innovations of such improvised methods firmly established the organic link between pedology, mineralogy, taxonomy and edaphology of five pedogenetically important soil orders like Alfisols, Mollisols, Ultisols, Vertisols and Inceptisols (Pal et al. 2012, 2014; Srivastava et al. 2015). Scope of innovations in improvised methods could be accomplished during the last decade and a half, when the focus of soil research changed qualitatively due to the use of high resolution mineralogical, micromorphological and age-control tools along with the geomorphic and climatic history. This further helped to measure the relatively subtler processes related to pedology of the present and past geological periods, and also to enhance the knowledge base on mineralogical transformations and their impact in rationalizing the soil taxonomic database(Pal et al. 2012, 2014; Srivastava et al. 2015). Such basic and fundamental information on major soil types of the country provided unique guiding principles to develop several simple methods to identify the important pedogenic processes, mineral transformation, taxonomic rationale and solving some queer edaphological issues of major soil types of India. Some of the developed simple methods are well tested in a recent robust National Agricultural Innovation Project (Georeferenced Soil Information System for Land Use Planning and Monitoring Soil and Land Quality for Agriculture, Bhattacharyya and Pal 2014). The worthiness of simple methods is demonstrated in the following few chapters. The usefulness of such methods in unravelling many interesting pedological, edaphological, mineralogical and taxonomical issues of soils of the country has been well established. More than one-third of the soils of the world are tropical soils and thus adoption of simple and ingenious methods by both students and young soil researchers to study pedology and edaphology of Indian tropical soils and for similar soils elsewhere may help in an expeditious development of national soil information system for enhancing crop productivity and maintaining soil health in the twenty-first century.

References

Bhattacharjee JC, Roychaudhury C, Landey RJ, Pandey S (1982) Bioclimatic analysis of India. NBSSLUP Bulletin 7, National bureau of soil survey and land use planning (ICAR), Nagpur

Bhattacharyya, T, Pal DK (Guest Editors) (2014) Special section on 'Georeferenced soil information system for land use planning and monitoring soil and land quality for agriculture'. Curr Sci 107:1400–1564

Bhattacharyya T, Pal DK, Deshpande SB (1993) Genesis and transformation of minerals in the formation of red (Alfisols) and black (Inceptisols and Vertisols) soils on Deccan basalt in the Western Ghats, India. J Soil Sci 44:159–171

Bhattacharyya T, Pal DK, Srivastava P (1999) Role of zeolites in persistence of high altitude ferruginous Alfisols of the humid tropical Western Ghats, India. Geoderma 90:263–276

Bhattacharyya T, Pal DK, Srivastava P (2000) Formation of gibbsite in presence of 2:1 minerals: an example from Ultisols of Northeast India. Clay Miner 35:827–840

Bhattacharyya T, Pal DK, Chandran P, Ray SK (2005) Land-use, clay mineral type and organic carbon content in two Mollisols–Alfisols–Vertisols catenary sequences of tropical India. Clay Res 24:105–122

Bhattacharyya T, Pal DK, Lal S, Chandran P, Ray SK (2006) Formation and persistence of Mollisols on zeolitic Deccan basalt of humid tropical India. Geoderma 146:609–620

Bhattacharyya T, Pal DK, Chandran P, Ray SK, Mandal C, Telpande B (2008) Soil carbon storage capacity as a tool to prioritise areas for carbon sequestration. Curr Sci 95:482–494

Bhattacharyya T, Sarkar D, Sehgal J, Velayutham M, Gajbhiye KS, Nagar AP, Nimkhedkar SS (2009) Soil taxonomic database of India and the states (1:250,000 scale), vol 143. NBSS&LUP Publication, Nagpur, p 266

Bhattacharyya T, Chandran P, Ray SK, Mandal C, Tiwary P, Pal DK, Wani SP, Sahrawat KL (2014) Processes determining the sequestration and maintenance of carbon in soils: a synthesis of research from tropical India. In: Soil horizon., Published July 9, 2014, pp 1–16. https://doi.org/10.2136/sh14-01-0001

Chandran P, Ray SK, Bhattacharyya T, Srivastava P, Krishnan P, Pal DK (2005) Lateritic soils of Kerala, India: their mineralogy, genesis and taxonomy. Aust J of Soil Res 43:839–852

Eswaran H, Kimble J, Cook T, Beinroth FH (1992) Soil diversity in the tropics: implications for agricultural development. In: Lal R, Sanchez PA (eds) Myths and science of soils of the tropics. SSSA special publication number29. SSSA, Inc and ACA, Inc, Madison, pp 1–16

Pal DK (2017) A treatise of Indian and tropical soils. Springer International Publishing AG, Cham

Pal DK, Deshpande SB, Venugopal KR, Kalbande AR (1989) Formation of di- and trioctahedral smectite as evidence for paleo-climatic changes in southern and central peninsular India. Geoderma 45:175–184

Pal DK, Kalbande AR, Deshpande SB, Sehgal JL (1994) Evidence of clay illuviation in sodic soils of indo-Gangetic plain since the Holocene. Soil Sci 158:465–473

Pal DK, Srivastava P, Bhattacharyya T (2003) Clay illuviation in calcareous soils of the semi-arid part of the indo-Gangetic Plains, India. Geoderma115:177–192

Pal DK, Bhattacharyya T, Chandran P, Ray SK, Satyavathi PLA, Durge SL, Raja P, Maurya UK (2009) Vertisols (cracking clay soils) in a climosequence of peninsular India: evidence for Holocene climate changes. Quatern Int 209:6–21

Pal DK, Wani SP, Sahrawat KL (2012) Vertisols of tropical Indian environments: pedology and edaphology. Geoderma 189-190:28–49

Pal DK, Wani SP, Sahrawat KL, Srivastava P (2014) Red ferruginous soils of tropical Indian environments: a review of the pedogenic processes and its implications for edaphology. Catena 121:260–278. https://doi.org/10.1016/j.catena2014.05.023

References

Pal DK, Wani SP, Sahrawat KL (2015) Carbon sequestration in Indian soils: present status and the potential. Proc Natl Acad Sci Biol Sci (NASB) 85:337–358. https://doi.org/10.1007/s40011-014-0351-6

Sanchez PA (1976) Properties and management of soils in the tropics. Wiley, New York

Srivastava P, Pal DK, Aruche KM, Wani SP, Sahrawat KL (2015) Soils of the indo-Gangetic Plains: a pedogenic response to landscape stability, climatic variability and anthropogenic activity during the Holocene. Earth-Sci Rev 140:54–71. https://doi.org/10.1016/j.earscirev.2014.10.010

Chapter 2
Evidence of Clay Illuviation in Soils of the Indo-Gangetic Alluvial Plains (IGP) and Red Ferruginous (RF) Soils

Abstract In the US system of soil classification, specific criteria are detailed out to define objectively the minimum evidence of clay illuviation required for an argillic horizon, which are particle size distribution relative to an overlying horizon and either clay skins on ped surfaces or oriented clay occupying 1% or more of the cross section. These criteria are not infallible when applied in many soil types occurring in semi-arid (SAT) and humid (HT) tropical climates of India. Pedologists while working in the micaceous Indo-Gangetic Alluvial (IGP) soils of the north-western India have often experienced clay-enriched textural B-horizons however without the identifiable clay skins by a 10 x hand lens. On the other hand, some pedologists considered the textural B-horizons as argillic on the basis of increased clay and the presence of field-identifiable clay skins or void argillans (impure type) and some of them considered the clay enrichment due to the sedimentation processes, geogenic origins and in situ weathering of biotite particles.

Loamy to clayey Mollisols, Alfisols and Ultisols of HT climate, which are in general mild to strongly acidic, have clay enriched B horizons and maintain the required base saturation but the identification of the argillic horizons in Ultisols is still not a straight forward criterion. The introduction of 'Kandic' concept in the US Soil Taxonomy ignores the requirement for argillic horizons in Ultisols. Although many Ultisols of the HT parts of southern peninsular area and north-east hill (NEH) regions qualify for Kandic horizon, scientific explanation is still awaited to address the pedogenic processes that are responsible for the clay enriched B-horizons but without the field identifiable clay skins. Identification of clay skins in the field is a tricky issue and soil micro-micromorphological thin section studies indicate the presence of either impure clay pedofeatures or less strongly oriented void argillans with low birefringence, which do not satisfy the basic requirement of pure void argillans as stipulated by the US Taxonomy. Therefore to circumvent this predicament demonstration of simple but scientifically sound analytical methods to ensure the formation of the argillic horizon through clay illuviation process are described in this chapter as a step towards precise and unambiguous definitions of soil taxa.

Keywords Clay illuviation · Argillic horizon · IGP soils · RF soils · SAT and HT climates

2.1 Introduction

The process of clay eluviation-illuviation is recognized at a high level in soil classification (Soil Survey Staff 1975; 1999) and in the legend of the Soil Map of the World (FAO/UNESCO 1974). Translocation of clay is considered to be an important process in pedogenesis; in addition, accessory properties of soils with argillic horizon are significant to the understanding of landscapes and the use of soils (Soil Survey Staff 1975). In soil classification, specific criteria are given that are intended to define objectively the minimum evidence of clay illuviation required for an argillic horizon. Major criteria of an argillic horizon are particle size distribution relative to an overlying horizon and either clay skins on ped surfaces or oriented clay occupying 1% or more of the cross section. On many occasions, pedologists while working in the micaceous IGP soils of the north-western (NW) India have experienced clay-enriched textural B-horizons but found no identifiable clay skins by a 10 x hand lens. However, some of them considered the textural B-horizons as argillic on the basis of increased clay and the presence of field-identifiable clay skins (Bhargava et al. 1981; Manchanda and Khanna 1981; Pal and Bhargava 1985; Tomar 1987) or void argillans (Karale et al. 1974; Sehgal et al. 1975). Clay enrichment has also been contributed to sedimentation processes (Hurelbrink and Fehrenbacher 1970), geogenic origins (Kooistra 1982), in situ weathering of biotite particles (Manchanda et al. 1983).

A dire need was felt at national level to resolve the diverse understanding on the genesis of textural B-horizons that exist with and without identifiable clay skins. Few recent comprehensive findings by Pal et al. (1994, 2003) on benchmark soils of the NW part of the IGP in the states of Haryana and Uttar Pradesh focused on the homogeneity of the parent material, degree of preferential movement of clay, clay mineral transformation, clay enrichment by differential weathering and sedimentation processes, and qualitative and quantitative aspects of void argillans. These studies indicated that (i) the parent material and the mineralogy of the clay fractions are uniform, (ii) the clay in the B-horizons is not formed in place, (iii) the fine clay(<0.2 μm) resulting substantially from weathering of biotite has preferentially translocated downwards causing an apparent decrease of clay mica with depth, (iv) the void argillans are typically of the type 'impure' clay pedofeatures and (v) the decrease in the clay mica (<2 μm) with depth could be a sure test of clay illuviation even when clay skins identifiable in the field are absent. It is realized by many pedologists that the identification of clay skins in field is a tricky feature, and in the event of missing such observation and record, a major pedogenic history in geological time scale in a stable landscape will certainly be undermined if such soils with clay enriched B-horizons are classed as Inceptisols. Criteria of clay skins and pure clay pedofeatures (in terms of pure void argillans) required to recognize argillic horizon, become incompatible in most of the tropical Indian soils developed under both SAT and HT climates. While identification of clay skins in the field is a tricky issue, the requirement of pure clay pedofeatures is not satisfied by the presence of either impure clay pedofeatures or less strongly oriented void argillans with low birefringence. Some strong pedogenic explanations are provided in support of these observations, which would add further knowledge to those hitherto documented for the less well oriented void argillans (Pal et al. 1994, 2003). It is now understood that while the presence of clay skins and pure clay pedofeatures is a clear evidence of illuviation, the absence of clay skins and presence of either impure clay pedofeatures or weak to moderately oriented void argillans with low birefringence, does not necessarily mean that there has been no illuviation of clay particles. To resolve this problem, application of micro-morphological thin section studies for void argillans and /or semi-quantification of clay minerals by X-ray diffraction (XRD) method are of much help as demonstrated by Pal et al. (1994, 2003). However, such sophisticated instrumental help, though provide enough guiding principles in the subject matter, and are still not easily accessible to most of the national soil laboratories. To circumvent this predicament urgency is felt to demonstrate simple but scientifically sound analytical methods to support the clay illuviation processes in Indian tropical soils, which are described in this chapter.

2.2 Depth Function of Clay Mica: An Incontrovertible Evidence of Clay Illuviation in IGP Soils

Two benchmark soils of the NW part of the IGP, one Typic Natrustalfs (Karnal district, Haryana state) and one Typic Haplustalf (Sitapur district, Uttar Pradesh), which have requisite criteria of particle size distribution, clay skins and > 1% void argillans (as the 'impure type' of clay pedofeatures) of the cross section, were subjected to semi-quantitative (SQ) estimates of the clay minerals in <2 µm clay fractions by XRD method (Table 2.1). The SQ content of <2 µm mica decreases with pedon depth due to the preferential movement of fine clay vermiculite and smectite from the A horizon to the B horizon. This decrease is illusory because it shows an increase with depth when clay mica is expressed on a fine earth basis (Glenn et al. 1960; Kapoor et al. 1981; Pal and Bhargava 1985; Pal et al. 1994) (Table 2.1), suggesting this decrease is a sure test of clay illuviation.

2.2.1 Development of a Simple Analytical Method

In the event of lack of the above instrumental facilities, a simple analytical protocol proposed by Pal (1997), insists on the quantitative determination of <2 µm clay mica of each soil horizon of the pedon, by digesting the <2 µm clay fractions (obtained during the particle size determination by the standard pipette method after removing the cementing agents) with $HF-HClO_4$ for determining K content. Then the K value is to be converted to K_2O content to the percentage of mica following the equation, 10% K_2O is equal to 100% mica (Kiely and Jackson 1965). Quantitative

Table 2.1 Semi-quantitative and quantitative estimates of the clay minerals of representative IGP soils

Horizon	Depth cm	M[a]	Clay minerals in <2 μm clay fractions (%) ML	V	S	Ch	K	M[b]%	Clay mica on fine earth basis % SQ[c]	Q[d]
Karnal: Typic Natrustalfs										
A1	0–6	46	22	7	12	6	9	51	5.6	6.1
A3	6–18	42	21	9	Trace	5	7	49	7.6	8.8
Bt21	18–49	36	16	Trace	2	12	6	48	8.0	10.5
Bt22	49–74	34	15	Trace	5	8	5	46	8.0	10.8
IIB31	74–136	37	19	Trace	Trace	11	16	37	7.0	7.0
Sitapur: Typic Haplustalf										
Ap	0–26	40	9	13	32	6	Trace	45	12.7	7.2
B1	26–42	34	12	13	31	Trace	10	40	13.2	12.9
Bt21	42–66	36	14	11	34	Trace	5	40	13.9	13.2
Bt22	66–94	30	12	13	40	Trace	5	38	12.5	12.6
B3	94–110	29	19	15	32	Trace	5	30	10.7	10.4

[a]M = Mica; ML = 1.0–1.4 nm minerals; V = Vermiculite; S = Smectite: Ch = Chlorite; Trace <5%
[b]M = Quantitative clay mica content (Kiely and Jackson 1965)
[c]SQ = Semi-quantitative mica content
[d]Q = Quantitative mica content (Adapted from Pal 1997)

value of clay mica shows similar depth trend on clay basis and also on fine earth basis as observed with SQ value (Table 2.1), which strongly suggests that the decreasing depth trend of quantitative value of clay mica is also a sure test of clay illuviation even when clay skins are not identified in the field, and such soils need to be placed in Alfisol order (Pal et al. 1994) as a step towards precise and unambiguous definitions of soil taxa.

2.3 Clay Illuviation in Red Ferruginous (RF) Soils of the Humid Tropical (HT) vis-a-vis IGP Soils of Semi-Arid (SAT) Climate

Mollisols, Alfisols and Ultisols, which are in general mild to strongly acidic, have clay enriched B horizons, and their texture varies from loamy to clayey (Bhattacharyya et al. 2009). These soils maintain the required base saturation because of the enrichment of basic cations of the parent materials derived from calc-gneiss and limestone but the identification of the argillic horizons in Ultisols is still not a straight forward criterion in the field like in many other parts of HT world. This predicament often confuses pedologists while placing soils in appropriate soil orders of the US Soil Taxonomy (Pal et al. 2014).

The introduction of 'Kandic' concept in Soil Taxonomy (Soil Survey Staff 1990), however, advises to ignore the requirement for argillic horizons in Ultisols.

Following this criterion, many Ultisols of the HT parts of southern peninsular area and north-east hill (NEH) regions now qualify for Kandic horizon as they meet its sub-surface textural and depth requirement (Pal et al. 2014). But no explanation is yet available to address the pedogenic processes that are responsible for the clay enriched B-horizons that meet the criteria of an argillic horizon but lack in field identifiable clay skins in low-activity clay (LAC) soils like Ultisols. However, many non-LAC soils of the NEH are developed in uniform parent materials and have requisite clay enrichment for the argillic horizon but without identifiable clay skins in the field (Sen et al. 1997). A similar observation is also reported in SAT soils of the IGP (Pal et al. 1994). In the absence of confirmation of clay illuviation process by thin section studies, such acid soils of NEH with clay enriched B horizons have been hitherto placed under Dystrochrepts, which undermines a major pedogenic process operating in these soils (Sen et al. 1997). Therefore, to investigate the precise cause-effect relation of the presence/absence of clay skins (that are generally identified by a 10 x hand lens in the field) a detailed micro-morphological study of soils of HT climate vis-a- vis the IGP soils of SAT environments, is warranted.

2.4 Micro-Morphological Thin Section Studies to Resolve the Argillic Horizon Issue

According to the US Soil Taxonomy (Soil Survey Staff 1975), the clay illuviation process is described as follows. The clay particles move in a suspended state. When water is absorbed by the dry peds, the ped faces or void walls act as a filter thereby retaining the clay particles that are deposited in the sub-surface horizons. These platelets are oriented parallel to the surface of deposition, which result in optically oriented pure void argillans through the process of clay illuviation (Fig. 2.1a- as text book reference, Bullock et al. 1985). The clay platelets remain in face to face association or parallel, or oriented aggregation (Fig. 2.1b, c) when the flocculation of the clay suspension is not induced by the presence of salts (Van Olphen 1966). In reality, the IGP soils of SAT climate with moderate to high sodicity do not show the presence of pure void argillans (Fig. 2.2a) (Pal et al. 1994, 2003; Srivastava et al. 2015, 2016). Even in slight to moderate sodic environment of carbonates and bicarbonates of Na^+ ions, deflocculation of clay suspension would disengage face to face association of clay platelets (Van Olphen 1966) and consequently lead to impairment of parallel orientation clay platelets (Fig. 2.2b, c). In this colloidal state, translocation of the fine clay particles would result in textural pedofeatures of the 'impure' type as observed in Natrustalfs and Haplustalfs of the IGP in SAT environment (Fig. 2.2a) (Pal et al. 1994, 2003). This demonstrates a specific pedogenic process that is different than those described so far for the genesis of less oriented void argillans (Brewer 1972; Bullock and Thompson 1985), and the 'impure clay pedofeatures' remain as a typical pedofeatures in SAT soils of the IGP (Pal et al. 1994, 2003; Srivastava et al. 2015, 2016).

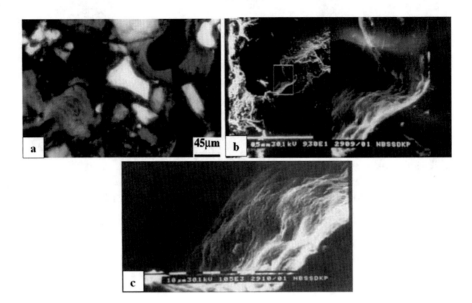

Fig. 2.1 Representative optical photomicrograph in cross-polarized light of (**a**) strongly oriented void argillans showing micro-laminations, (**b**) full view of illuvial clay pedofeatures along a void under scanning electron microscope (SEM) showing strong orientation of fine clay coating in the void, (**c**) expanded part of the same showing strong parallel lamination of fine clay particles of ferruginous soils (Rhodustalfs) of southern India (Adapted from Venugopal et al. 1991; Pal et al. 2003)

In acidic Alfisols, Mollisols and Ultisols of HT climate the void argillans are expected to be 'pure' type (Fig. 2.1a) in absence of any salts. However, limited data on the clay pedofeatures in such soils in India (Kooistra 1982) and elsewhere (Eswaran 1972) indicate the limited presence of pure void argillans with high birefringence (Fig. 2.1a). But in acidic Mollisols and Alfisols of HT climate of the Western Ghats, the presence of pure void argillans is rarely observed. Instead, clay pedofeatures with thin continuous clay coatings to strongly oriented micro-laminated coatings along the voids are very common (Fig. 2.3a, b); however with less birefringence (Lal 2000) as compared to that of pure void argillans shown in Fig. 2.1a. But such clay pedofeatures are not considered to be the result of current pedogenic processes in acid soils of HT climate (Eswaran 1972; Kooistra 1982). The dispersion of clay particles is possible under slightly acidic to moderately alkaline pH conditions at a very low electrolyte concentration that ensures a pH higher than that required for the zero point of charge for complete dispersion of clay (Eswaran and Sys 1979). Therefore, dispersion of clay particles and its subsequent translocation and accumulation in the B horizons occurs at the initial stage of soil formation when the pH is moderately alkaline and remains above the point of zero charge. This specific chemical environment would result in deflocculation that disengages face-to-face association of clay particles, and finally impairs the parallel orientation of clay particles (Van Olphen 1966). In this colloidal state, translocation

2.4 Micro-Morphological Thin Section Studies to Resolve the Argillic Horizon Issue

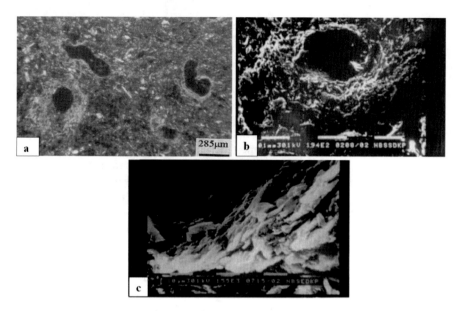

Fig. 2.2 Representative thin-section photographs showing typical micromorphological cross-polarized light of (**a**) impure clay pedofeatures, (**b**) the same under scanning electron microscope (SEM), (**c**) expanded part of the same under SEM showing poorly oriented clay platelets in soils (Haplustalfs) of the NW part of the IGP (Adapted from Pal et al. 2003)

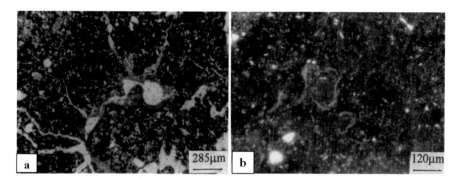

Fig. 2.3 Representative thin-section photographs of typical clay pedofeatures in cross-polarized light showing strongly oriented micro-laminated coating along voids by dotted clay of the acidic Mollisols (**a**) and clay pedofeatures showing thin continuous clay coatings along voids of acidic Alfisols (**b**) of the Western Ghats (Adapted from Lal 2000)

of the fine clay particles to the sub-surface horizons and accumulation thereof, would result in 'impure clay pedofeatures' (Fig. 2.2a) as observed in the IGP soils of SAT environment (Pal et al. 1994, 2003; Srivastava et al. 2015, 2016). However, the presence of void argillans observed in acidic Mollisols, Alfisols and Ultisols of HT climate suggests that with the advancement in weathering and leaching of bases, and concomitant lowering of pH during the initial stages of pedogenesis in HT climate, clay platelets could remain moderately in face to face association or parallel, or oriented aggregates when the flocculation of the clay suspension was not induced by the presence of salts. This justifies the statement of Kooistra (1982) and Eswaran 1972) that the clay illuviation in acidic soils of HT climate is not a current pedogenetic process (Pal et al. 2014). Fine clay fractions of the Mollisols and Alfisols of the Western Ghats, the Ultisols of Kerala, and the Ultisols of Meghalaya of the NEH region have dominant portion of kaolin (not a pure kaolinite but a 0.7 nm mineral, which is an interstratified 1.4–0.7 nm clay minerals) (Bhattacharyya et al. 1993, 1999, 2000, 2006; Chandran et al. 2005). The void argillans of these soils are not very strongly laminated and also lack strong birefringence (Fig. 2.3a, b) in contrast to pure void argillans (Fig. 2.1a). The soils developed in HT climate with pure void argillans (Venugopalan et al. 1991), contain fairly well crystalline fine clay (<0.2 μm) kaolinite mineral with higher order reflections (Fig. 2.4a), which shows hexagonal in shape with clear-cut edges under scanning electron microscope (Fig. 2.4b). It appears that the fine clay kaolin of soils of the Western Ghats, Kerala and NEH, being an interstratified 0.7 nm mineral (Fig. 2.4c), fails to orient strongly as parallel clay platelets to each other like pure fine clay kaolinite while getting accumulated in the B-horizons even under a prolonged pedogenesis in salt free chemical environment of HT climate (Pal et al. 2014). It is, however, quite likely that these types of clay pedofeatures may not yield the right kind of clay skins that can be easily identified in the field unambiguously by the pedologists.

2.4.1 Simple Analytical Way to Recognize the Argillic horizon in Indian Tropical Soils

In view of the above facts, it is now clear that the US Soil Taxonomy criteria of clay skins and pure clay pedofeatures (in terms of pure void argillans) to recognize argillic horizon, become incompatible in most of the tropical Indian soils developed under SAT and HT climates. While identification of clay skins in the field is not straightforward, the requirement of pure clay pedofeatures is not satisfied by the presence of either impure clay pedofeatures or less strongly oriented void argillans with low birefringence. Some strong pedogenic explanations are provided in support of these observations, which would add further knowledge to those hitherto documented for the less well oriented void argillans (Pal et al. 1994, 2003). This suggests that the absence of clay skins and presence of either impure clay pedofeatures or weak to moderately oriented void argillans with low birefringence also indicate the presence of argillic horizon through an active clay illuviation process.

In soils of the HT climate, the enrichment of clay in the B horizons through illuviation of clay particles did occur in the uniform parent material (generally ascertained by the absence of any lithological discontinuity during the morphological examination of the soil profiles in the field, supported by the clay free sand /silt ratio

2.4 Micro-Morphological Thin Section Studies to Resolve the Argillic Horizon Issue

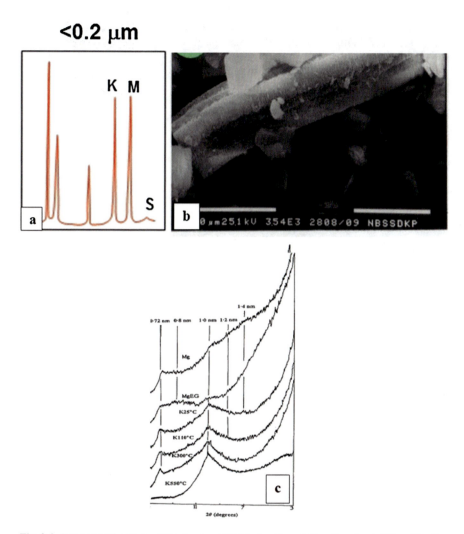

Fig. 2.4 Representative X-ray diffractogram (XRD) of well crystalline fine clay (<0.2 μm) kaolinite mineral (K) with higher order reflections (**a**), hexagonal kaolinite with clear-cut edges under scanning electron microscope of SAT Alfisols of Mysore plateau of Karnataka (**b**), and representative XRD of fine clay (<0.2 μm) 0.72 nm mineral of Alfisols of HT climate (Adapted from Pal et al. 1989; Pal and Sarma 2002; Bhattacharyya et al. 1993)

of each soil horizon, Pal and Deshpande 1987), and thus discounts the other possibilities of the clay accumulation in the B horizons.

In view of the clay illuviation in the uniform parent material on a stable geomorphic surface of the landscape as the major pedogenic process for a prolonged geological period under HT (Pal et al. 2014) and SAT (Srivastava et al. 2015, 2016) climates; and the fulfillment of the textural requirement of a Bt horizon for loam to clay loam soils, it would be prudent to waive the clay skins criterion in recognizing the presence of argillic horizon in these soils. This simple tool may prove to be useful in establishing the inherent relationships between environmental factors and properties that essentially reflect the genesis of argillic horizon in similar situations. In addition, it will help in applying the precise and unambiguous definitions of soil taxa.

References

Bhargava GP, Pal DK, Kapoor BS, Goswami SC (1981) Characteristics and genesis of some sodic soils in the indo-Gangetic alluvial plain of Haryana and Uttar Pradesh. J. Indian Soc Soil Sci 29:61–70

Bhattacharyya T, Pal DK, Deshpande SB (1993) Genesis and transformation of minerals in the formation of red (Alfisols) and black (Inceptisols and Vertisols) soils on Deccan basalt in the western Ghats, India. J Soil Sci 44:159–171

Bhattacharyya T, Pal DK, Srivastava P (1999) Role of zeolites in persistence of high altitude ferruginous Alfisols of the humid tropical western Ghats, India. Geoderma 90:263–276

Bhattacharyya T, Pal DK, Srivastava P (2000) Formation of gibbsite in presence of 2:1 minerals: an example from Ultisols of Northeast India. Clay Miner 35:827–840

Bhattacharyya T, Pal DK, Lal S, Chandran P, Ray SK (2006) Formation and persistence of Mollisols on Zeolitic Deccan basalt of humid tropical India. Geoderma 136:609–620

Bhattacharyya T, Sarkar D, Sehgal JL, Velayutham M, Gajbhiye KS, Nagar AP, Nimkhedkar SS (2009) Soil taxonomic database of India and the states (1:250,000 scale), vol 143. NBSSLUP, Publication, p 266

Brewer R (1972) The basis of interpretation of soil micromorphological data. Geoderma 8:81–94

Bullock P, Thompson ML (1985) Micromorphology of Alfisols. In: Douglas LA, Thompson ML (eds) Soil micromorphology and soil classification. Soil Science Society of America, Madison, pp 17–47

Bullock P, Fedoroff N, Jongerious A, Stoops G, Tursina T (1985) Handbook of soil thin section description. Waine Research Publication, p 152

Chandran P, Ray SK, Bhattacharyya T, Srivastava P, Krishnan P, Pal DK (2005) Lateritic soils of Kerala, India: their mineralogy, genesis and taxonomy. Aust J Soil Res 43:839–852

Eswaran H (1972) Micromorphological indicators of pedogenesis in some tropical soils derived from basalts from Nicaragua. Geoderma 7:15–31

Eswaran H, Sys C (1979) Argillic horizon in LAC soils formation and significance to classification. Pédologie 29:175–190

FAO/UNESCO (1974) Soil map of the world, vol 1. Legend Published by FAO, Rome

Hurelbrink RL, Fehrenbacher JB (1970) Soils and stratigraphy of a portion of the Gola River fan of Uttar Pradesh. Soil Sci Soc Am Proc 37:911–916

Glenn RC, Jackson ML, Hole FD, Lee GB (1960) Chemical weathering of layer silicate clays in loess-derived Tama silt loam of South-Western Wisconsin. Clay Clay Miner 8:63–83

Kapoor BS, Singh HB, Goswami SC (1981) Distribution of illite in some alluvial soils of the indo-Gangetic plain. J Indian Soc Soil Sci 29:572–574

Karale RL, Bisdom EBA, Jongerious J (1974) Micromorphological studies on diagnostic subsurface horizons of some alluvial soils in the Meerut district of Uttar Pradesh. J Indian Soc Soil Sci 22:70–76

Kiely PV, Jackson ML (1965) Quartz, feldspar, and mica determination for soils by sodium pyrosulphate fusion. Soil Sci Soc Am Proc 29:159–163

References

Kooistra MJ (1982) Micromorphological analysis and characterization of 70 benchmark soils of India. Soil survey institute, Wageningen

Lal S (2000) Characteristics, genesis and use potential of soils of the Western Ghats, Maharashtra. Ph. D Thesis, Dr. P.D.V.K., Akola, Maharashtra, India

Manchanda ML, Khanna SS (1981) Soil salinity and landscape relationships in part of Haryana state. J Indian Soc Soil Sci 29:493–503

Manchanda ML, Khanna SS, Garalapuri VN (1983) Weathering dispersibility and clay skins in subsurface diagnostic horizons of soils in parts of Haryana. J Indian Soc Soil Sci 31:565–571

Pal DK (1997) An improvised method to identify clay illuviation in soils of indo-Gangetic plains. Clay Res 16:46–50

Pal DK, Bhargava GP (1985) Clay illuviation in a sodic soil of the northwestern part of the indo-Gangetic alluvial plain. Clay Res 4:7–13

Pal DK, Deshpande SB (1987) Parent material, mineralogy and genesis of two benchmark soils of Kashmir valley. J Indian Soil Sci 35:690–698

Pal DK, Deshpande SB, Venugopal KR, Kalbande AR (1989) Formation of di- and trioctahedral smectite as an evidence for paleoclimatic changes in southern and central peninsular India. Geoderma 45:175–184

Pal DK, Kalbande AR, Deshpande SB, Sehgal JL (1994) Evidence of clay illuviation in sodic soils of north-western part of the indo-Gangetic Plains since the Holocene. Soil Sci 158:465–473

Pal DK, Sarma VAK (2002) Chemical composition of soils. In: Fundamentals of soil science. Indian Society of Soil Science, New Delhi, pp 209–227

Pal DK, Srivastava P, Bhattacharyya T (2003) Clay illuviation in calcareous soils of the semi-arid part of the indo-Gangetic Plains, India. Geoderma 115:177–192

Pal DK, Wani SP, Sahrawat KL, Srivastava P (2014) Red ferruginous soils of tropical Indian environments: a review of the pedogenic processes and its implications for edaphology. Catena 121:260–278. https://doi.org/10.1016/j.catena2014.05.023

Sehgal JL, Hall GF, Bhargava GP (1975) An appraisal of the problems in classifying saline-sodic soils of the indo-Gangetic plain in NW India. Geoderma 14:75–91

Sen TK, Nayak DC, Singh RS, Dubey PN, Maji AK, Chamuah GS, Sehgal JL (1997) Pedology and edaphology of benchmark acid soils of North-Eastern India. J Indian Soc Soil Sci 45:782–790

Soil Survey Staff (1975) Soil taxonomy: a basic system of soil classification for making and interpreting soil surveys. Agriculture handbook. Soil Conservation Service, US Dept. of Agriculture, Washington, DC, p 436

Soil Survey Staff (1990) Keys to soil taxonomy, 4th edn, 19. SMSS Technical Monograph, Blacksburg

Soil Survey Staff (1999) Soil taxonomy: a basic system of soil classification for making and interpreting soil surveys, USDA-SCS agricultural handbook no 436, 2nd edn. U.S. Govt Printing Office, Washington, DC

Srivastava P, Pal DK, Aruche KM, Wani SP, Sahrawat KL (2015) Soils of the indo-Gangetic Plains: a pedogenic response to landscape stability, climatic variability and anthropogenic activity during the Holocene. Earth-Sci Rev 140:54–71. https://doi.org/10.1016/j.earscirev.2014.10.010

Srivastava P, Aruche M, Arya A, Pal DK, Singh LP (2016) A micromorphological record of contemporary and relict pedogenic processes in soils of the indo-Gangetic Plains: implications for mineral weathering, provenance and climatic changes. Earth Surf Process Landf 41:771–790. https://doi.org/10.1002/esp.3862

Tomar KP (1987) Chemistry of pedogenesis in indo-Gangetic alluvial plains. J Indian Soc Soil Sci 38:405–414

Van Olphen H (1966) An introduction of clay colloid chemistry. Interscience, New York

Venugopal KR, Deshpande SB, Kalbande AR, Sehgal JL (1991) Textural pedofeatures (clay coatings) in a ferruginous soil from Bangalore plateau. Clay Res 10:30–35

Chapter 3
Clay Illuviation and Pedoturbation in SAT Vertisols

Abstract For a long time the apparent uniform distribution of clay throughout Vertisols was considered to be effect of haploidisation within the pedon caused by pedoturbation and in some cases the observed gradual increase in clay content with depth was thought to be due to inheritance from parent material. Recent systematic pedological studies on Vertisols that have no stratification in the parent material and no clay skins, indicated that their Bss horizons contain clay even up to 20%; an increase from the eluvial horizon. Such depth distribution of clay is due to clay illuviation process was confirmed by micro-morphological investigation of the thin sections, which indicated the presence of >2% impure clay pedofeatures. Thus the clay enriched Bss horizons in Vertisols suggests that pedoturbation was too much favoured as an important pedogenic process in Vertisols by the past researchers till early nineties, who envisaged pedoturbation would obliterate all evidence of illuviation. But in reality, pedoturbation in Vertisols is only a partially functional process, which cannot overshadow the more significant long-term clay illuviation process. Although the micro-morphological study of soil thin sections is a unique analytical tool to confirm clay illuviation process, for many of the national soil science laboratories it is truly a very distant facility. In its absence some simple analytical data are of much help in ensuring the clay illuviation process with certainty as major pedogenic process in Indian Vertisols, which are described in this chapter.

Keywords Tropical Vertisols · Clay illuviation · Pedoturbation

3.1 Introduction

For a long time it was conceived that the apparent uniform depth distribution of clay in Vertisols (Ahmad 1983; Murthy et al. 1982) is the result of haploidisation within the pedon that caused considerable pedoturbation (Mermut et al. 1996).However, some earlier studies reported that there is a gradual increase in clay content with depth (Dudal 1965), which was thought to be due to inheritance from the parent material only (Ahmad 1983). Studies on Vertisols by the Indian Council of Agricultural Research-National Bureau of Soil Survey and Land Use Planning

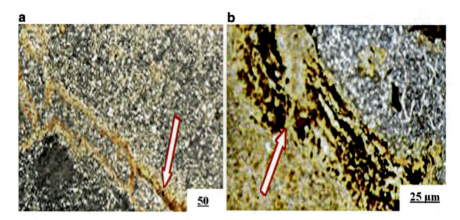

Fig. 3.1 Representative photographs of clay pedofeatures in cross-polarized light (**a**) Impure clay pedofeatures of SHM Vertisols. (**b**) Disrupted clay pedofeatures of SHD Vertisols (Adapted from Pal et al. 2009)

(ICAR-NBSS&LUP), Nagpur, India over the past decade and a half, indicated that (i) the clay content of the Bss horizons ranges from zero to substantially enriched with clay (~20% increase from the eluvial horizon; Pal et al. 2009, 2012), and (ii) morphological examination of the Vertisols indicated uniformity in the parent material and showed no clay skins. However, micro-morphological investigation of the thin sections indicated the presence of >2% impure clay pedofeatures (Fig. 3.1a), which are formed by the pedogenic process similar to that explained for the IGP soils of SAT environment (Pal et al. 1994, 2003). The impure clay pedofeatures have hitherto been considered as the sure test of clay illuviation in tropical Indian soils (Pal et al. 1994, 2003, 2009) and thus their presence confirms that the clay is enriched in the Bss horizons of Vertisols by clay illuviation. Clay illuviation was also identified in clay soils with vertic properties in Canada, Uruguay and Argentina (Pal et al. 2012). In earlier studies, it was thought that pedoturbation would obliterate all evidence of illuviation, except in the lower horizons (Eswaran et al. 1988; Mermut et al. 1996). This led Johnson et al. (1987) to consider this process to be an example of proisotropic pedoturbation caused by argilli-turbation. The pedoturbation was thought to destroy horizons or soil genetic layers and to make Vertisols revert to a simpler state. On the contrary, the clay-enrichment of the Bss horizons via illuviation suggests that the argilli-turbation is not a primary pedogenetic process in Vertisols (Pal et al. 2009) and represents a proanisotropism in the soil profile (Pal et al. 2012). The concept of proisotropic pedoturbation is also not compatible with the steady decrease in soil organic carbon, exchangeable Ca/Mg and by increases in total and fine clays, exchangeable sodium percentage (ESP), water-dispersible clay (WDC) and carbonate clay (fine earth based) with depth (Fig. 3.2) (Pal et al. 2009). Therefore, pedoturbation in Vertisols is a partially functional process and is unable to overshadow the more important long-term clay illuviation process.

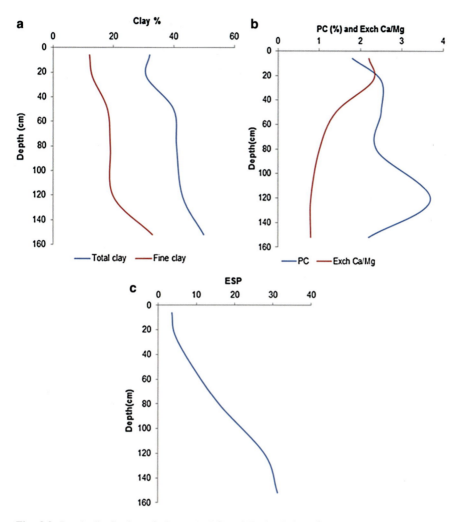

Fig. 3.2 Depth distribution of clays (**a**), PC and Exch. Ca/Mg (**b**) and ESP (**c**) in a Sodic Calciusterts (Adapted from Pal et al. 2009)

3.2 Uniqueness of Clay Illuviation in Highly Smectitic Vertisols

Smectite parent material from the weathering Deccan basalt has been deposited in the lower topographic positions during a previous wetter climate and the Vertisols were developed in such alluvium during the drier climate of the Holocene period (Pal et al. 2009). A study of the Vertisols of the sub-humid, semi-arid and arid climates of Peninsular India clearly indicates that both plagioclase and micas are either fresh or weakly to moderately altered, suggesting that chemical weathering of these minerals has not been substantial (Pal et al. 2012). These data discount the formation of smectite during the development of Vertisols (Srivastava et al. 2002) and validate the hypothesis that Vertisol formation reflects a positive entropy change (Smeck et al. 1983). This is in contrast to the formation of fine clay vermiculite and smectite at the expense of biotite mica in IGP soils under SAT environment and also to the weathering of vermiculite and smectite towards the formation of kaolin in RF soils of HT climate. Therefore, it is necessary to highlight the factors that cause the dispersion of fine smectitic parent material, its translocation to and accumulation in sub-surface horizons.

The majorities of Vertisols are calcareous and contain a considerable amount of water dispersible clay (WDC), which increases with depth (Pal et al. 2009, 2012). Such depth function suggests that the dispersion of clay smectite is possible under slightly acidic to moderately alkaline pH conditions at a very low electrolyte concentration (ECe $\leq$ 1 me L^{-1}; Pal et al. 2009, 2012) that ensures a pH higher than the zero point of charge required for a full dispersion of clay (Eswaran and Sys 1979). Previous studies postulated that carbonate removal is a pre-requisite for illuviation of clay, as Ca^{2+} ions enhance the flocculation and immobilization of colloidal material (Bartelli and Odell 1960); however, low quantities of soluble Ca^{2+} ions ($\ll$5me L^{-1}, Pal et al. 2009, 2012) are not enough to cause flocculation of clay particles in Vertisols. Therefore, movement of deflocculated fine clay smectite (and its subsequent accumulation in the Bss horizons) is possible in non-calcareous as well as calcareous Vertisols (Pal et al. 2003). It is interesting, however, to note that despite being calcareous, Vertisols have low concentration of soluble Ca-ions but the exchange sites are dominated by Ca-ions followed by Mg.

Amidst a restricted weathering of plagioclase, the primary source of Ca^{2+} ions in the soil solution is the dissolution of non-pedogenic CaCO$_3$ (NPCs) (Fig. 3.3a) (Srivastava et al. 2002). The precipitation of CaCO$_3$ as pedogenic CaCO$_3$ (PC) in the form of diffuse nodules and dense aggregates of micrite crystals (Fig. 3.3b) is generally observed in close proximity to the NPCs (Fig. 3.3c) and enhances the pH and the relative abundance of Na$^+$ ions in soil exchange and in solution. The Na$^+$ions in turn cause dispersion of clay smectites, and the dispersed smectites translocate even in the presence of CaCO$_3$. The formation of PC facilitates the deflocculation of clay particles and their subsequent movement downward. This pedogenic processes suggest that PC formation and clay illuviation are two concurrent and contemporary pedogenic events (Pal et al. 2003, 2009, 2012). The translocation of the Na-saturated fine clays causes however, a decreasing depth distribution of exchangeable calcium percentage (ECP), saturated hydraulic conductivity (sHC), and an increasing depth distribution of water dispersible clay (WDC), exchangeable magnesium percentage (EMP), exchangeable sodium percentage (ESP), PC (carbonate clay on fine-earth basis) in the majority of SAT Vertisols of India (Pal et al. 2009, 2012).

3.3 Simple Tool to Establish Clay Illuviation as the Primary and Pedoturbation...

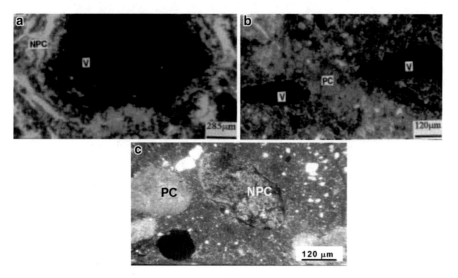

Fig. 3.3 Micromorphological features of both PC and NPC in the Vertisols of central, western and southern peninsular India, in cross-polarized light (**a**) NPC showing dissolution and removal of CaCO$_3$ from the nodule (V). (**b**) PC as diffuse nodules and dense aggregates of micrite crystals in the groundmass. (**c**) PCs are generally observed in close proximity to the NPCs, the distance could be ≤3 μm (Adapted from Pal et al. 2000; Srivastava et al. 2002)

3.3 Simple Tool to Establish Clay Illuviation as the Primary and Pedoturbation as Partially Functional Pedogenic Process in Vertisols

The clay illuviation in Vertisols of tropical Indian environment is reflected as 'impure clay pedofeatures' instead of pure void argillans, under thin section studies and clay illuviation could not yield enough clay skins identifiable in the field. This suggests that the requirement of the clay skins or the pure void argillans as criteria for the Bt horizon of the US Soil Taxonomy, is not fulfilled, except however, the textural requirement for clayey soils. For a long time an opinion was held that pedoturbation in Vertisols would normally prevent the accumulation of clay as illuviation cutans or destroy developed cutans beyond recognition (Blokhuis 1982). It is however interesting to observe the ubiquitous presence of 'clay pedofeatures' in Vertisols and soils with very high shrink-swell properties (Pal et al. 2009, 2012; Srivastava et al. 2010, 2015, 2016). In addition to these observations, presence of disrupted clay pedofeatures (Fig. 3.1b) are also documented amidst the abundant presence of clay pedofeatures (Pal et al. 2009), suggesting there has been no role of pedoturbation in destroying the illuvial clay pedofeatures. Therefore, pedoturbation in Vertisols is a partially functional process. Although argillic horizons are common in Vertisols, the Bt horizon does not get better than their dominant property (slickensides) because Vertisol soil order keys out before the Alfisols according to the US

Soil Taxonomy. Thus, at present, these soils would still be classed as Vertisols (Pal et al. 2009, 2012).

In view of the above discussion, it is clear that Vertisols are developed in smectitic parent clays under SAT climatic environment of the Holocene period. The formation of PC primarily causes the clay illuviation process, which is the major pedogenic process in the uniform smectitic clay parent material, and causes the clay increase in the sub-surface horizons, pH, clay carbonate, EMP, ESP and reduction in sHC, and thus discounts the pedoturbation as the major functional pedogenic process. In light of these basic pedogenic processes, the requirement of clay skins and pure void argillans for confirming the clay illuviation process in Vertisols appears to be of secondary importance. Therefore, some of the above key properties or a combination of them (especially the increasing depth distribution of total and fine clays, EMP, ESP and clay $CaCO_3$ on fine earth basis, and by the decrease in exchangeable Ca/Mg)(Fig. 3.2) would be enough to recognize the clay illuviation in Indian Vertisols and elsewhere for a precise and unambiguous definition of soil taxa, especially when the facility for soil thin section studies is not in place.

References

Ahmad N (1983) Vertisols, pedogenesis and soil taxonomy. In: Wilding LP, Smeck NE, Hall GF (eds) The soil orders, vol II. Elsevier, Amsterdam, pp 91–123

Bartelli LJ, Odell RT (1960) Laboratory studies and genesis of a clay-enriched horizon in the lowest part of the solum of some brunizem and gray-brown podzolic soils in Illinois. Soil Sci Soc Amer Proc 24:390–395

Blokhuis AA (1982) Morphology and genesis of Vertisols. Vertisols and rice soils in the tropics: Transactions 12th international congress of soil science, New Delhi, pp 23–47

Dudal R (1965) Dark clay soils of tropical and subtropical regions. FAO Agric. Dev. Paper, 83. FAO, Rome, p 161

Eswaran H, Sys C (1979) Argillic horizon in LAC soils formation and significance to classification. Pédologie 29:175–190

Eswaran H, Kimble J, Cook T (1988) Properties, genesis and classification of Vertisols. Classification, management and use potential of swell–shrink soils. In: Hirekerur LR, Pal DK, Sehgal JL, Deshpande SB (eds) INWOSS, October 24–28, 1988. National Bureau of Soil Survey and Land Use Planning, Nagpur, pp 1–22

Johnson DL, Watson-Stegner D, Johnson DN, Schaetzl RJ (1987) Proisotropic and proanisotropic processes of pedoturbation. Soil Sci 143:278–292

Mermut AR, Padmanabham E, Eswaran H, Dasog GS (1996) Pedogenesis. In: Ahmad N, Mermut AR (eds) Vertisols and technologies for their management. Elsevier, Amsterdam, pp 43–61

Murthy RS, Bhattacharjee JC, Landey RJ, Pofali RM (1982) Distribution, characteristics and classification of Vertisols. Vertisols and rice soils of the tropics, Symposia paper II, 12th International congress of soil science, New Delhi. Indian Society of Soil Science, pp 3–22

Pal DK, Kalbande AR, Deshpande SB, Sehgal JL (1994) Evidence of clay illuviation in sodic soils of north-western part of the indo-Gangetic Plains since the Holocene. Soil Sci 158:465–473

Pal DK, Dasog GS, Vadivelu S, Ahuja RL, Bhattacharyya T (2000) Secondary calcium carbonate in soils of arid and semi-arid regions of India. In: Lal R, Kimble JM, Eswaran H, Stewart BA (eds) Global climate change and pedogenic carbonates. Lewis Publishers, Boca Raton, pp 149–185

References

Pal DK, Srivastava P, Bhattacharyya T (2003) Clay illuviation in calcareous soils of the semi-arid part of the indo-Gangetic Plains, India. Geoderma 115:177–192

Pal DK, Bhattacharyya T, Chandran P, Ray SK, Satyavathi PLA, Durge SL, Raja P, Maurya UK (2009) Vertisols (cracking clay soils) in a climosequence of peninsular India: evidence for Holocene climate changes. Quatern Int 209:6–21

Pal DK, Wani SP, Sahrawat KL (2012) Vertisols of tropical Indian environments: pedology and edaphology. Geoderma 189–190:28–49

Smeck NE, Runge ECA, Mackintosh EE (1983) Dynamics and genetic modeling of soil system. In: Wilding LP, Smeck NE, Hall GF (eds) Pedogenesis and soil taxonomy: 1. Concepts and interactions. Elsevier, New York, pp 51–81

Srivastava P, Bhattacharyya T, Pal DK (2002) Significance of the formation of calcium carbonate minerals in the pedogenesis and management of cracking clay soils (Vertisols) of India. Clay Clay Miner 50:111–126

Srivastava P, Rajak MK, Sinha R, Pal DK, Bhattacharyya T (2010) A high resolution micromorphological record of the late quaternary paleosols from ganga-Yamuna interfluve: stratigraphic and paleoclimatic implications. Quatern Int 227:127–142

Srivastava P, Pal DK, Aruche KM, Wani SP, Sahrawat KL (2015) Soils of the indo-Gangetic Plains: a pedogenic response to landscape stability, climatic variability and anthropogenic activity during the Holocene. Earth Sci Rev 140:54–71. https://doi.org/10.1016/j.earscirev.2014.10.010

Srivastava P, Aruche M, Arya A, Pal DK, Singh LP (2016) A micromorphological record of contemporary and relict pedogenic processes in soils of the Indo-Gangetic Plains: implications for mineral weathering, provenance and climatic changes. Earth Surf Proc Land 41:771–790. https://doi.org/10.1002/esp.3862

Chapter 4
Cracking Depths in Indian Vertisols: Evidence of Holocene Climate Change

Abstract In Indian sub-continent, Vertisols in humid tropical (HT), sub-humid moist (SHM), sub-humid dry (SHD), semi-arid moist (SAM), semi-arid dry (SAD) and arid dry (AD) climatic environments indicate their occurrence in a climosequence. The soils show a change in their morphological, physical, chemical and micromorphological properties due to change of climate from humid to arid during the Holocene period. Formation of pedogenic $CaCO_3$ (PC), illuviation of clay and the development of subsoil sodicity are concurrent, contemporary and active pedogenetic processes operating during the climate change of the Holocene period. These processes impaired the hydraulic properties of soils in general, and in soils of drier climates in particular. As a result, cracking pattern, chemical composition and plasmic fabric were more modified in soils of the drier climates. Such modifications in soil properties have a place in the rationale of Vertisol order of the US Soil Taxonomy. The soils of wetter climates (HT, SHM and SHD) are grouped in Typic Haplusterts whereas the soils of drier climates (SAM, SAD and AD) are classified as Aridic Haplusterts, Sodic Haplusterts and Sodic Calciusterts. Such pedological study demonstrates how the depth of cracking in Vertisols in a climosequence can be used as simple analytical tools in inferring the change in climate in a geologic period.

Keywords Tropical Vertisols · Climosequence · Cracking depths · Holocene climate change

In response to the global climatic events during the Quaternary, the soils in many parts of the world witnessed climatic fluctuations, especially in the last post-glacial period. During the Quaternary, climate changes have also been frequent (Ritter 1986). In major parts of the Indian subcontinent, climate change from humid to semi-arid did occur during the late Holocene (Pal et al. 2012, 2014; Srivastava et al. 2015). Due to this climate shift, major soil types of India are becoming calcareous and sodic as a result of regressive pedogenesis under the present SAT environments (Pal et al. 2013). The regressive pedogenesis has caused modifications in the physical and chemical properties of soils. Such unfavourable soil properties reduce the possibility of successful growing of winter crops (Pal et al. 2009).

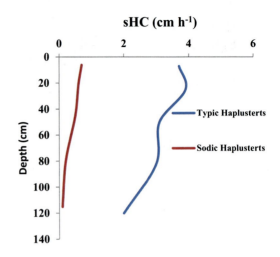

Fig. 4.1 Saturated hydraulic conductivity (sHC/hr) of the representative non-sodic (Typic Haplusterts) and sodic (Sodic Haplusterts) Vertisols (Adapted from Pal et al. 2009)

In the Indian sub-continent, majority of the Vertisols occur in the lower piedmont plains or valleys (Pal and Deshpande 1987) or in micro depressions (Bhattacharyya et al. 1993). They are developed mainly in the alluvium of weathering Deccan basalt deposited during previous wetter climate (Pal and Deshpande 1987; Bhattacharyya et al. 1993), during the Holocene period (Pal et al. 2001, 2009). They occur in humid tropical (HT), sub-humid moist (SHM), sub-humid dry (SHD), semi-arid moist (SAM), semi-arid dry (SAD) and arid dry (AD) climatic environments and thus indicate an array of Vertisols in a climosequence. Such occurrence of Vertisols suggests the influence of basaltic parent materials to form similar soils under different climatic conditions (Mohr et al. 1972). Although all these soils are grouped under Vertisol order (Soil Survey Staff 2006), the soils show a change in their morphological, physical, chemical and micromorphological properties in the climosequence.

Soils of HT climate are dominated by Ca^{2+} ions in their exchange complex throughout depth. However, in the sub-humid climates Mg^{2+} ions tend to dominate in the lower horizons. Due to the typical regressive pedogenesis operative in the SAT soils, the sub-humid moist to aridic climatic environments caused a progressive formation of pedogenic calcium carbonates (PC) with the concomitant increase in Na^+ ions in soil solution. This facilitates the translocation of Na-clay in the soil profile, which causes the increase in pH, decrease in Ca/Mg ratio of exchange sites with depth and finally in the development of subsoil sodicity. The reduction in mean annual rainfall (MAR) from sub-humid moist to arid climates accelerated the formation of PC and thus the soils of semi-arid and arid climates (SAM, SAD and AD) are more calcareous and sodic than soils of other climates (SHM and SHD).

Formation of PC, illuviation of clay and the development of subsoil sodicity are concurrent, contemporary and active pedogenetic processes operating during the climate change of the Holocene period. These processes impaired the hydraulic conductivity (HC) of soils in general, and in soils of drier climates in particular. The HC decreased rapidly with depth in all the soils from Typic to Sodic Haplusterts (Fig. 4.1) but the decrease was sharper in Sodic Haplusterts of drier climate (SAM, SAD, and AD) because of their subsoil sodicity (ESP > 5 < 15, Pal et al. 2009). The

4 Cracking Depths in Indian Vertisols: Evidence of Holocene Climate Change

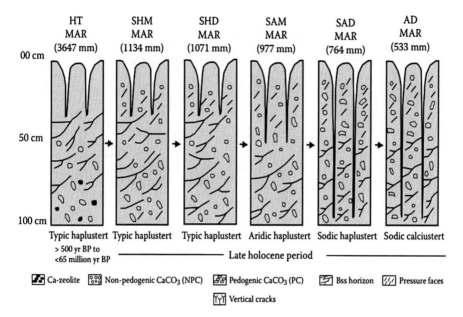

Fig. 4.2 Successive stages of pedogenic evolution of Vertisols in a climosequence (Adapted from Pal et al. 2009) (MAR = mean annual rainfall; HT = humid tropical; SHM = sub- humid moist; SHD = sub- humid dry; SAM = semi -arid moist; SAD = semi-arid dry and AD = arid dry)

decreased HC restricts vertical and lateral movement of water in the subsoils. During the very hot summer months, this would result in much less water in the subsoil of the drier climates (Pal et al. 2001). This is evident from the deep cracks (> 0.5 cm) cutting through the Bss horizons in Vertisols of drier climates (Fig. 4.2) associated with the formation of calcareous and alkaline soils. Generally, cracks in non-sodic Vertisols (Typic Haplusterts) of wetter climate extend down to the slickenside zones (Figs. 4.2a, 4.3a, Pal et al. 2001) and show strong plasma separation (Fig. 4.4a). However in Vertisols with subsoil sodicity, the lack of adequate soil water during the shrink–swell cycles restricts the swelling of smectite and results in weaker plasma separation in the soils (Sodic Haplusterts) of the drier climates (Fig. 4.4b) (Pal et al. 2009, 2012). As a result, cracking pattern, chemical composition and plasmic fabric were more modified in soils of the drier climates. Such modifications in soil properties have a place in the rationale of Vertisol order of the US Soil Taxonomy. The soils of wetter climates (HT, SHM and SHD) are grouped in Typic Haplusterts whereas the soils of drier climates (SAM, SAD and AD) are classified as Aridic Haplusterts, Sodic Haplusterts and Sodic Calciusterts. Therefore, the exact soil grouping using the US Soil Taxonomy to decode behaviour of climate and its change proves that soils have a tremendous memory and store the past episodes carefully. An inherent pedogenetic relationship exist between the dry climate, formation of PC, exchangeable Ca/Mg, ESP and HC (Kadu et al. 2003), and the reduced HC ultimately cause the deepening of the cracks, down the pedon depth. Thus, the appearance of cracks cutting through the Bss horizons may be an easy and

Signature of climate change

Fig. 4.3 Cracks in non-sodic Vertisols (Typic Haplusterts) extend down to the slickenside zones only (**a**) but they puncture the whole slickensided horizons in Sodic Haplusterts/Calciusterts (**b**) - a sign of climate change causing subsoil sodicity and reduced soil productivity (Photos, courtesy-author)

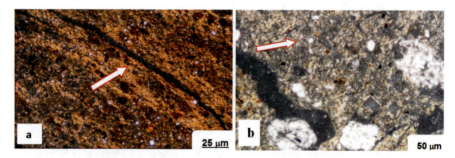

Fig. 4.4 Representative photographs of plasmic fabric in cross polarized light. (**a**) Strong parallel oriented plasmic fabric in Typic Haplusterts of wetter climate and (**b**) weak plasma separation as mosaic/stippled-speckled plasmic fabric in Sodic Haplusterts of dry climates (Adapted from Pal et al. 2009)

simple tool (Fig. 4.3b) to infer the climate change from humid to dry climates in Vertisols not only in the Indian semiarid tropics, but also in similar climatic areas elsewhere.

References

Bhattacharyya T, Pal DK, Deshpande SB (1993) Genesis and transformation of minerals in the formation of red (Alfisols) and black (Inceptisols and Vertisols) soils on Deccan basalt in the western Ghats, India. J Soil Sci 44:159–171

Kadu PR, Vaidya PH, Balpande SS, Satyavathi PLA, Pal DK (2003) Use of hydraulic conductivity to evaluate the suitability of Vertisols for deep-rooted crops in semi-arid parts of Central India. Soil Use Manag 19:208–216

Mohr ECJ, Van Baren FA, Van Schuylenborgh J (1972) Tropical soils- a comprehensive study of their genesis. Mouton-Ichtiarbaru-Van Hoeve, The Hague

Pal DK, Deshpande SB (1987) Characteristics and genesis of minerals in some benchmark Vertisols of India. Pedologie (Ghent) 37:259–275

Pal DK, Balpande SS, Srivastava P (2001) Polygenetic Vertisols of the Purna Valley of Central India. Catena 43:231–249

Pal DK, Bhattacharyya T, Chandran P, Ray SK, Satyavathi PLA, Durge SL, Raja P, Maurya UK (2009) Vertisols (cracking clay soils) in a climosequence of peninsular India: evidence for Holocene climate changes. Quatern Int 209:6–21

Pal DK, Wani SP, Sahrawat KL (2012) Vertisols of tropical Indian environments: pedology and edaphology. Geoderma 189-190:28–49

Pal DK, Sarkar D, Bhattacharyya T, Datta SC, Chandran P, Ray SK (2013) Impact of climate change in soils of semi-arid tropics (SAT). In: Bhattacharyya et al (eds) Climate change and agriculture. Studium Press, New Delhi, pp 113–121

Pal DK, Wani SP, Sahrawat KL, Srivastava P (2014) Red ferruginous soils of tropical Indian environments: a review of the pedogenic processes and its implications for edaphology. Catena 121:260–278. https://doi.org/10.1016/j.catena2014.05.023

Ritter DF (1986) Process geomorphology. Wm C Brown, Dubuque Iowa, p 603

Soil Survey Staff (2006) Keys to soil taxonomy, 10th edn. United States Department of Agriculture, Natural Resources Conservation Service, Washington, DC

Srivastava P, Pal DK, Aruche KM, Wani SP, Sahrawat KL (2015) Soils of the indo-Gangetic Plains: a pedogenic response to landscape stability, climatic variability and anthropogenic activity during the Holocene. Earth-Sci Rev 140:54–71. https://doi.org/10.1016/j.earscirev.2014.10.010

Chapter 5
Unique Depth Distribution of Clays in SAT Alfisols: Evidence of Landscape Modifications

Abstract Many SAT Alfisols of the Indian sub-continent show an increase in clays with depth to a maximum and then decreases until it remains constant or completely disappears and fulfills the textural criterion of an Alfisol. In contrast, many SAT red ferruginous Alfisols, mainly Paleustalfs and Rhodustalfs of southern India have clay content of about 10–15% in the Ap horizon, immediately followed by a well-developed argillic (Bt) horizon with a clay content of >30%. This is a unique situation in pedological parlance and thus needed a scientific explanation. Detailed clay mineralogical investigations along with the geomorphic and climatic history of such Alfisols indicate that these Alfisols are developed on the old rock system of the earth and represent relict paleosols. The unique depth distribution of clays bears the testimony of the landscape modification affected through the truncation of Alfisols developed in the preceding humid climate.

Keywords Alfisols · Southern India · Truncation of soil profile · Climate change

To ascertain the clay illuviation in soils, it is prudent to determine the particle size distribution adapting the international pipette method after removal of organic matter, $CaCO_3$ and free Fe oxides. Sand (2000-50 µm), silt (50-2 µm), total clay (<2 µm) and fine clay (<0.2 µm) fractions are then separated according to the procedure of Jackson (1979). In many SAT Alfisols of the Indian sub-continent, clays increase with depth. The ratio of clay content in the clay-rich Bt horizon compared to that of the A horizon is >1.2, which suggests that these soils are fairly well developed. In such soils, the clay increases with depth to a maximum and then decreases until it remains constant or completely disappears. Illuviation of clay particles usually results in the development of clay skins that can be recognized in the field with a 10X lens (Pal et al. 1994, 2003). In contrast, many SAT red ferruginous Alfisols, mainly Paleustalfs and Rhodustalfs (Natarajan et al. 1997; Shiva Prasad et al. 1998) have clay content of about 10–15% in the Ap horizon, immediately followed by a well-developed argillic (Bt) horizon with a clay content of >30%. In the Bt3 or Bt4 horizons at a depth of about 60–70 cm, clay distribution shows in general, a decreasing trend (Fig. 5.1). The thin Ap horizon is a disturbed horizon due to soil erosion or ploughing. If this horizon is ignored, then the depth distribution of clay in such

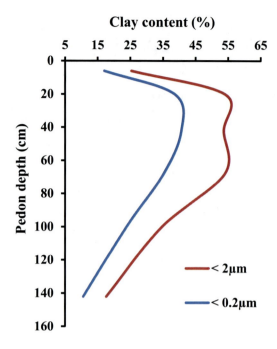

Fig. 5.1 A unique depth-wise distribution of total clay (<2 μm) and fine clay (< 0.2 μm) fractions with pedon depth of a representative SAT Alfisol having the Ap (0-12 cm), B21t (12-29 cm), B22t (29–56 cm), B23t (56–81 cm), B24t (81–118 cm) and BC (118–167) horizons (Adapted from Pal and Deshpande 1987)

Alfisols shows an increasing trend up in the profile, which is generally seen in a juvenile residual soil profile. However, these Alfisols are developed on the old rock system of the earth and represent relict paleosols (Pal et al. 1989, 2014).

In parts of southern and eastern India, laterite mounds and laterite plateau remnants are scattered over the landscape (Pal and Roy 1978; Rengasamy et al. 1978; Subramanian and Mani 1981). In central and western India thin-to-thick (25–300 cm) laterite cappings occur on various rock types ranging in ages from Archean to Gondwanas (Pascoe 1965; Sahasrabudhe and Deshmukh 1981; Subramanian and Mani 1981). Large numbers of massive granitic tors in gneissic terrain bear the evidence of exhumation (Pal and Deshpande 1986; Pal and Deshpande 1987) during the dry period following prolonged deep weathering in the HT climate that prevailed from the Upper Cretaceous (Subramanian and Mani 1981) until Plio-Pleistocene. The Plio-Pleistocene was a transition period when the climate became drier with rising of the Western Ghats (Brunner 1970). As a result, the upper layers of RF soils (mainly Paleustalfs and Rhodustalfs, Pal et al. 1989; Natarajan et al. 1997; Shiva Prasad et al. 1998) formed in the preceding HT climate, were truncated by multiple arid erosional cycles (Chandran et al. 2000; Murali et al. 1978; Rengasamy et al. 1978). Due to probable truncation of the upper layers, the coarse and fine clay contents presently show an upward increase in the solum, a sharp decline in the Ap horizon (Fig. 5.1) and argillans immediately beneath the Ap horizon (Chandran et al. 2000; Murthy et al. 1982; Pal et al. 1989). Many such soils are at present mildly acidic and calcareous (Bhattacharyya et al. 2009), but often indicate the presence of with pure void argillans (Kooistra 1982; Venugopal et al. 1991;

5 Unique Depth Distribution of Clays in SAT Alfisols: Evidence of Landscape...

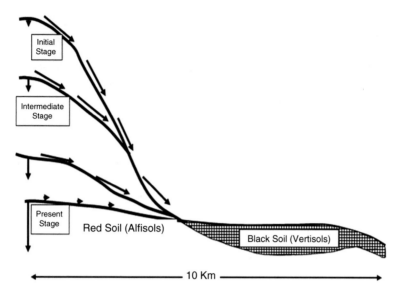

Fig. 5.2 Schematic diagram of the pedon site of red soils (Alfisols) and black soils, (Vertisols) showing the landscape reduction process explaining the formation of spatially associated red and black soils (Adapted from Pal 2008)

Venugopal 1997), which is however is not the result of clay illuviation as a current pedogenetic process in such SAT Alfisols (Eswaran 1972; Kooistra 1982). The presence of pure void argillans reported in some RF soils suggests that with the advancement in weathering and leaching of bases, and concomitant lowering of pH during the initial stages of pedogenesis in previous HT climate, clay platelets could remain in face to face association or parallel, or oriented aggregation (See Chap. 2, Fig. 2.1a) when the flocculation of the clay suspension was not induced by the presence of salts (Van Van Olphen 1966), particularly the carbonates and bicarbonates of sodium (Pal et al. 1994). Therefore, in such SAT Alfisols, which are relict paleosols (Pal et al. 1989), modification in the geomorphic surface is also evident from the presence of broken argillans (papules) (Kooistra 1982; Venugopal 1997; Venugopal et al. 1991) in the soil solum. Such broken argillans /papules represent low-energy short distance transport in the upland undulating areas (Srivastava et al. 2010). Such landscape reduction process in the past and present has created a unique spatially associated red-black soil complex (Fig. 5.2) in many parts of southern and western India (Pal et al. 2012, 2014; Bhattacharyya et al. 1993; Pal 2008; Rengasamy et al. 1978).

Most of the SAT Alfisols have fine loamy to clayey texture (Bhattacharyya et al. 2009), and are in general calcareous and almost neutral to mildly alkaline in nature. These Alfisols are spatially associated with Vertisols (Typic/Aridic/Sodic Haplusterts, Pal et al. 2012) and are dominated by kaolin with small amount of smectite (Pal and Deshpande 1987). In some SAT Alfisols, it was observed that kaolin decreases and smectite increases with depth, and the sum of these two minerals are fairly constant from the B to the C (saprolites) horizon (Pal et al. 1989). Such relation suggests the transformation of smectite(Sm) to kaolin (Kl). Present SAT climatic conditions are not considered severe enough for transformation and weathering to kaolinite and also for the formation of huge amount of smectite in ferruginous and spatially associated Vertisols (Bhattacharyya et al. 1993; Pal et al. 2012). Dioctahedral smectite was thus the first weathering product of peninsular gneiss during the HT climate of pre-Pliocene period, but its stability was ephemeral in this climate as evident from its transformation to kaolin. Both kaolin (Sm–Kl) and smectite could be preserved in such soils because of termination of HT climate during the Plio-Pleistocene transition (Pal et al. 1989). The spatially associated SAT Vertisols in the lower topographic positions are developed during the dry climate of the Holocene period in the deposited smectitic parent material of the previous much wetter climate (Pal et al. 2009). The preservation of the crystallinity of SAT Vertisols' smectite in depressions and the lack of transformation of primary minerals validate the hypothesis of positive entropy change during the formation of Vertisols (Srivastava et al. 2002). Therefore, the formation of such Vertisols also helps to infer climate change in geographical areas covered by the SAT Alfisols in the Peninsular India (Pal et al. 2012). SAT Alfisols of the Peninsular India overlying the saprolites of metamorphic rocks, dominated either by kaolin or dioctahedral smectite, are thus relict paleosols (Pal et al. 1989). These relict soils have been affected by the climatic change from humid to semi-arid (SA) climatic conditions during the Plio-Pleistocene transition period. During SAT climate trioctahedral smectite was formed from the sand and silt size biotite, which survived earlier weathering during the HT climate (Chandran et al. 2000; Pal et al. 1989). In addition, selected such soils show the presence of pedogenic calcium carbonates (PC) as spongy nodules and cluster of carnonets needles below 40 cm depth, indicating their formation in the prevailing SAT environments (Pal et al. 2014). During the SAT climate for millions of years many such soils are at present mildly acidic and calcareous (Bhattacharyya et al. 2009), and their clay enriched B horizons often indicate the presence of impure clay pedofeatures (Venugopal 1997) along with pure void argillans (Kooistra 1982; Venugopal et al. 1991; Venugopal 1997). In the SAT climate, the precipitation of $CaCO_3$ facilitates the deflocculation of clay particles and their subsequent translocation and accumulation in the Bt horizons (Pal et al. 2003), but in the presence of various types of other illuvial pedofeatures (Venugopal 1997), it is probably imprudent to suggest a precise pedogenic process for the formation of poorly oriented (impure) clay pedofeatures (Pal et al. 2014), especially when SAT Alfisols still show an upward increase of clay in the profile (after ignoring the thin Ap horizon) (Fig. 5.1). This unique depth trend could be related well to the landscape modification affected through the truncation of Alfisols developed in the preceding humid climate. Therefore, landscape modifications in terms of landscape reduction process (Bhattacharyya et al. 1993; Pal 1988) explains well the formation of spatially associated Alfisols and Vertisols under almost same topographical situation in SAT environments of southern India (Fig. 5.2) (Pal et al. 2012, 2014).

References

Bhattacharyya T, Pal DK, Deshpande SB (1993) Genesis and transformation of minerals in the formation of red (Alfisols) and black (Inceptisols and Vertisols) soils on Deccan basalt in the western Ghats, India. J Soil Sci 44:159–171

Bhattacharyya T, Sarkar D, Sehgal JL, Velayutham M, Gajbhiye KS, Nagar AP, Nimkhedkar SS (2009) Soil Taxonomic database of India and the states (1:250,000 scale), NBSSLUP Publ. 143, NBSS&LUP, Nagpur, India, p 266

Brunner H (1970) Pleistozäne Klimaschwankungen im Bereich des Östlichen Mysore Plateaus (Stüd Indien). Andean Geol 19:72–82

Chandran P, Ray SK, Bhattacharyya T, Krishnan P, Pal DK (2000) Clay minerals in two ferruginous soils of southern India. Clay Res 19:77–85

Eswaran H (1972) Micromorphological indicators of pedogenesis in some tropical soils derived from basalts from Nicaragua. Geoderma 7:15–31

Jackson ML (1979) Soil chemical analysis. Advanced course, 2nd edn. Published by the author University of Wisconsin, USA

Kooistra MJ (1982) Micromorphological analysis and characterization of 70 benchmark soils of India. Soil Survey Institute, Wageningen

Murali V, Krishnamurti GSR, Sarma VAK (1978) Clay mineral distribution in two topo sequences of tropical soils of India. Geoderma 20:257–269

Murthy RS, Hirekerur LR, Deshpande SB, Venkat Rao BV (eds) (1982) Benchmark soils of India. National Bureau of Soil Survey and Land Use Planning, Nagpur, p 374

Natarajan A, Reddy PSA, Sehgal J, Velayutham M (1997) Soil Resources of Tamil Nadu for land use planning. NBSS Publ. 46b (Soils of India Series), National Bureau of Soil Survey and Land Use Planning, Nagpur, India, 88 pp + 4 sheets of soil map on 1:500,000 scale

Pal DK (1988) On the formation of red and black soils in southern India. In: Hirekerur LR, Pal DK, Sehgal JL, Deshpande SB (eds) Transactions international workshop swell-shrink soils. Oxford and IBH, New Delhi, pp 81–82

Pal DK (2008) Soils and their mineral formation as tools in paleopedological and geomorphological studies. J Indian Soc Soil Sci 56:378–387

Pal DK, Deshpande SB (1986) Genesis and transformation of clay minerals of two red soil (Ustalf) pedons of South India. XIII Int Congr Soil Sci Trans 4:1471–1472

Pal DK, Deshpande SB (1987) Genesis of clay minerals in a red and black complex soils of southern India. Clay Res 6:6–13

Pal DK, Roy BB (1978) Characteristics and genesis of some red and lateritic soils occurring in topo sequence in eastern part of India. Indian Agric 22:9–28

Pal DK, Deshpande SB, Venugopal KR, Kalbande AR (1989) Formation of di and trioctahedral smectite as an evidence for paleoclimatic changes in southern and central peninsular India. Geoderma 45:175–184

Pal DK, Kalbande AR, Deshpande SB, Sehgal JL (1994) Evidence of clay illuviation in sodic soils of north-western part of the indo-Gangetic plains since the Holocene. Soil Sci 158:465–473

Pal DK, Srivastava P, Bhattacharyya T (2003) Clay illuviation in calcareous soils of the semi-arid part of the indo-Gangetic Plains, India. Geoderma 115:177–192

Pal DK, Bhattacharyya T, Chandran P, Ray SK, Satyavathi PLA, Durge SL, Raja P, Maurya UK (2009) Vertisols (cracking clay soils) in a climosequence of peninsular India: evidence for Holocene climate changes. Quatern Inter 209:6–21

Pal DK, Wani SP, Sahrawat KL (2012) Vertisols of tropical Indian environments: pedology and edaphology. Geoderma 189–190:28–49

Pal DK, Wani SP, Sahrawat KL, Srivastava P (2014) Red ferruginous soils of tropical Indian environments: a review of the pedogenic processes and its implications for edaphology. Catena 121:260–278. https://doi.org/10.1016/j.catena2014.05.023

Pascoe EH (1965) A manual of the geology of India and Burma. Manager of Publications, Delhi, p 2017

Rengasamy P, Sarma VAK, Murthy RS, Krishna Murthy GSR (1978) Mineralogy, genesis and classification of ferruginous soils of the eastern Mysore Plateau, India. J Soil Sci 29:431–445

Sahasrabudhe YS, Deshmukh SS (1981) The laterites of the Maharashtra state. In: Laterisation processes. Proc Inter Seminar. A.A. Balkema, Rotterdam, pp 209–220

Shiva Prasad CR, Reddy PSA, Sehgal J, Velayutham M (1998) Soils of Karnataka for optimising land use. NBSS Publ. 47b (Soils of India Series), National Bureau of Soil Survey and Land Use Planning, Nagpur, India, 111 pp + 4 sheets of soil map on 1:500,000 scale

Srivastava P, Bhattacharyya T, Pal DK (2002) Significance of the formation of calcium carbonate minerals in the pedogenesis and management of cracking clay soils (Vertisols) of India. Clay Clay Miner 50:111–126

Srivastava P, Rajak M, Sinha R, Pal DK, Bhattacharyya T (2010) A high resolution micromorphological record of the late quaternary Paleosols from ganga-Yamuna interfluve: stratigraphic and paleoclimatic implications. Quatern Int 227:127–142

Subramanian KS, Mani G (1981) Genetic and geomorphic aspects of laterites on high and low landforms in parts of Tamil Nadu, India. Laterisation processes. Proc Inter Seminar. A. A. Balkema, Rotterdam, pp 236–224

Van Olphen H (1966) An introduction of clay colloid chemistry. Interscience, New York

Venugopal KR (1997) Types of cutans in some ferruginous soils of Bangalore plateau and their relation with soil development. J Indian Soc Soil Sci 46:641–646

Venugopal KR, Deshpande SB, Kalbande AR, Sehgal JL (1991) Textural pedofeatures (clay coatings) in a ferruginous soil from Bangalore plateau. Clay Res 10:30–35

Chapter 6
Easy Identifications of Soil Modifiers

Abstract Although the presence of soil modifiers such as gypsum, calcium carbonate, palygorskite and zeolites, is being adequately reported in soils for the last two decades by the Indian soil and earth scientists, their unique role in the formation of soils in tropical humid climates, in re-defining in saline sodic soils, in land evaluation and impairing hydraulic properties, in fine-tuning the exiting soil classification scheme and also the management practices to enhance and sustain the productivity of Indian soils is recently realized. Use of sophisticated analytical instruments such as X-ray diffraction (XRD) technique, soil thin section studies and by observing the morphological features by scanning electron microscope (SEM) is required to identify the sand, silt and clay-sized soil modifiers. But such instrumental facilities are rarely available in national soil laboratories. In view of their immense impact in changing the pedo-chemical environment of soils, emergence of simple analytical methods became imperative. In this chapter, few simple but useful chemical methods, which can help planning better use and management of soils, are discussed.

Keywords Soil modifiers · Simple methods for identification · Soil management

6.1 Introduction

Since the early years of 1990, the unique role of soil modifiers (such as gypsum, pedogenic calcium carbonates, zeolites and palygorskite) in fine-tuning the soil classification scheme and also the existing management practices in enhancing and sustaining the soils' productivity, was realized by the Indian pedologists (Pal 2013). Soil is a complex system and thus sometimes it is difficult to detect and identify small quantities of soil modifiers. Therefore, special care needs to be taken while preparing them for subsequent mineralogical analyses. Sand, silt and clay-sized soil modifiers can be conveniently identified by X-ray diffraction (XRD) technique, soil thin section studies and by observing their morphological features by scanning electron microscope (SEM) after ascertaining their optical characters under optical microscope. Over the past two decades however, the focus of research has shifted from general pedology to mineralogical and micro-morphological and climate research. Such research endeavour on major soil types of tropical environment of India (Pal et al. 2012a, 2014; Bhattacharyya et al. 2013; Srivastava et al. 2015, 2016) has helped in resolving many enigmatic edaphological aspects and also provided important hints how to develop simple analytical tools to identify the presence of soil modifiers even in absence of sophisticated instruments. In view of their remarkable influence in changing the pedo-chemical environment of soils, which are essential for better use and management, application of such simple tools appears to be imperative. Few such tools are described below.

6.2 Gypsum and Its Significance in Redefining Saline-Sodic Soils and Managing SAT Vertisols

Limited experimental evidence indicates that when saline-alkali soils of the NW parts of the IGP (Indo-Gangetic Alluvial Plains) are leached, they do not turn out to be more sodic; rather their ESP (exchangeable sodium percentage) is reduced (Leffelaar and Sharma 1977; Abrol and Bhumbla 1978). This is in contrast to what has been defined for saline-alkali soils in the United States Salinity Laboratory (Richards 1954), which states that when soluble salts are leached with water with low Ca^{2+} ions these soils become alkali. Reduction in ESP can be caused only through the increased solubilities of naturally present $CaCO_3$ and gypsum (Dieleman 1963) but in saline-alkali environment displacement of exchangeable Na by Ca^{2+} from $CaCO_3$ is of doubtful significance (Yaalon 1958). During the field examination the presence of gypsum was not ascertained and also the laboratory data do not indicate the presence of gypsum (Goyal et al. 1974), even though Ca^{2+} and Mg^{2+} ions constitute as sub-dominant cations in soil saturation extract. Ca^{2+} and Mg^{2+} ions together are dominant cations in irrigation water (Leffelaar and Sharma 1977). Such dominance of divalent cations like Ca and Mg strongly suggests the presence of gypsum in parent material of the saline-alkaline soils undertaken by Leffelaar and Sharma (1977). Real saline-sodic soils of the IGP when leached either by highly sodic waters (Sharma and Mondal 1981) or rain water (Pal et al. 2010) show reduced water transmission rate. Therefore, saline-sodic soils with gypsum can be finally designated by saline soils as proper mapping unit for their pragmatic soil management (Pal 2013).

Dry climatic environment is the prime factor for natural soil chemical degradation processes that induce the formation of pedogenic $CaCO_3$ (PC) and the concomitant development of sodicity (Pal et al. 2000, 2016). Despite this possibility, selected Vertisols of the SAD (semi-arid dry) climate in southern India, though highly calcareous, are non-sodic and support the growth of crops such as cotton, pigeon pea and sorghum. This is a paradoxical situation. However, pedological research (mainly soil thin section studies) on such soils indicates the presence of gypsum in these soils even in presence of good amount of both PC and NPC (non-pedogenic $CaCO_3$).The soils have a sHC (saturated hydraulic conductivity) > 30 mm hr^{-1} despite the rapid formation of PC, unlike in the zeolitic Vertisols of the SAD climate (Pal et al. 2009a). Although the sustainability of crop productivity in these Vertisols depends on the gypsum stock, the present poor productivity of cotton (approximately 2 t ha^{-1}) may be enhanced by irrigation because the inherent gypsum would prevent water logging because of better drainage (Pal et al. 2009b).

Generally in the IGP soils of the semi-arid tropical (SAT) environments, Na$^+$ and Mg^{2+} ions dominate on the soil exchange complex and Na$^+$ ion is the dominant cation and Cl$^-$ the dominant anion in the saturation extract. In contrast, in SAT Vertisols with gypsum (Pal 2013), Ca^{2+} and Mg^{2+} ions are the dominant cations on exchange complex whereas in saturation extract Ca^{2+} ion is more than Mg and Na, and SO$_4$ is more than Cl and CO$_3$. Often soils containing gypsum have base saturation (BS) > 100 and sHC >> 20 mm hr^{-1}, which are the sure test of its presence in soils. BS of calcareous soils can be easily determined in the laboratory following standard methods for the determination of the soil's cation exchange capacity and exchangeable Na and K (Richards 1954) and exchangeable Ca and Mg by 1 N NaCl solution extraction method (Piper 1966; Pal et al. 2009c). The sHC can be determined using a constant head permeameter (Richards 1954). Therefore, regular record of the presence of gypsum in IGP soils and Vertisols of SAT environments is very crucial for their precise delineation in detailed soil maps for their proper agricultural land use plans.

6.3 Pedogenic Calcium Carbonate (PC) and Its Role in Land Evaluation

In SAT soils, calcareousness is caused by the presence of both pedogenic and non-pedogenic CaCO$_3$, but the pedogenic formation of PC is not a favourable chemical reaction for soil health because this creates unfavourable physical conditions, caused by concomitant development of exchangeable sodium per cent (ESP) (Pal et al. 2016). The presence of pedogenic PC that is distinguished from the pedorelict CaCO$_3$(NPC) by the soil thin section studies (Pal et al. 2000), is common in major soil types of India (alluvial soils of the Indo-Gangetic plains, red ferruginous soils and shrink-swell soils).The development of sodicity is not realized yet in the desert soils due to their sandy textural class, ensuring better leaching of bicarbonates; and thus PC is generally observed at greater depth. In the loamy and clayey textured soils, the leaching of bicarbonates is slow and thus both PC and sodicity develop in upper horizons (Pal et al. 2000). In IGP and red ferruginous soils of SAT environment, the presence of NPC is rare but in Vertisols both PC and NPC is observed (Pal et al. 2000).Thus, the SAT Vertisols exhibit the presence of PC in their thin sections (Pal et al. 2000; Srivastava et al. 2002; Vaidya and Pal 2002) and have carbonate clay, which on a fine earth basis, increases with pedon depth likewise the depth distribution of ESP. However, exchangeable Ca/Mg ratio and sHC show a reverse depth distribution (See Chap. 3, Fig. 3.2b, c). In their studies to identify the soil properties that influence the yield of deep rooted crops in Vertisols of central SAT India, Kadu et al. (2003) established the pedogenic relationship between SAT climate, carbonate clay, ESP, exchangeable Ca/Mg ratio and sHC (See Chap. 8, Table 8.1). They also established a fact that all these soil properties are the yield influencing factors. Thus the formation of PC is a major pedogenetic process in SAT Vertisols and its amount as carbonate clay is a more important soil requirement than total $CaCO_3$ in soils (Kadu et al. 2003). When the facilities for soil thin section studies are not available, carbonate clay (Shields and Meyer 1964) can be determined on the basis of the gravimetric loss of carbon dioxide using Collin's calcimeter.

6.4 Zeolites and their Role in Soil Formation and Management

Several reviews have been published about the occurrence and properties of zeolites in soils (Pal et al. 2013; Bhattacharyya et al. 2015). Zeolites have been reported in sodic soils of the NW parts of the IGP (Kapoor et al. 1981) and they also occur as secondary minerals in the Deccan flood basalts of the Western Ghats in Maharashtra, India (Bhattacharyya et al. 2015). Among the commonly occurring species of zeolites, analcime in sodic soils of the IGP (Kapoor et al. 1981) and heulandite is widely distributed in cracking clay soils developed in the alluvium of the weathering Deccan basalts (Bhattacharyya et al. 2015). Zeolites have the ability to hydrate and dehydrate reversibly and to exchange some of their constituent cations and thus, can influence the pedochemical environment during the formation of soils.

Over the past few decades, natural zeolites have been examined for a variety of agricultural and environmental applications (Pal et al. 2013). These applications can result in direct or indirect incorporation of natural zeolites into soils. There are, however, very few studies on the role of zeolites in soil environments in expanding the basic knowledge in pedology and edaphology, except for some pioneering work reported from India as stated below.

6.4.1 Analcime in IGP Soils

The influence of analcime in soils is generally realized in the over estimation of the ESP (> 100) and CEC values when standard methods (Richards 1954) are followed. Thus, such overestimated ESP was often understood in the past to be an error in the analytical procedure followed for determining the exchangeable cations and CEC. This misunderstanding kept the pedologists away for a long time from the reality about the presence of zeolite in soils. However, realizing the role of analcime in over estimation of ESP and CEC, Gupta et al. (1985) proposed a pragmatic analytical method that could eliminate the role of analcime to obtain the realistic value of gypsum requirement of sodic soils. Overestimated ESP value would however cost more for the chemical amendment like mined gypsum.

6.4.2 Heulandite in Soils of the Deccan Basalt Areas

Some of the pioneering research on the role of heulandites in Mollisols, Alfisols and Vertisols pertain to (1) persistence of high-altitude Alfisols, Mollisols and Vertisols of the humid tropical Deccan basalt areas of the central and western peninsular India (Bhattacharyya et al. 1993, Bhattacharyya et al. 1999, Bhattacharyya et al. 2006); (2) the role of zeolites in redefining the sodic soils (Pal et al. 2006), (3) mitigation of Holocene climate change by the zeolitic Sodic Haplusterts (Pal et al. 2006, 2012a, 2012b), and (4) establishment of the link between zeolite and adsorption and desorption behaviour of nutrients in soils (Pal et al. 2013) because zeolites are effective slow-release fertilizers and soil conditioners (Ming and Allen 2001).

6.4.3 Simple Analytical Method to Identify Zeolites in Soils

Ming and Dixon (1987) developed a CEC procedure to quantify clinoptilolite. The authors, however, felt that the procedure needs further modification to quantify zeolites other than clinoptilolite because in nature there are more than 60 zeolites, each with unique crystal structures, ion-sieving properties, cation selectivity and CEC. It is a fact that there is no selective method to quantify the heulandite in soils which have other clay minerals. But the qualitative presence of heulandite in soils developed in Deccan basalt alluvium could be made by Bhattacharyya et al. (1999) who determined the CEC and extractable bases that provide indications for the possible presence of zeolites in soils. These authors determined the CEC of acidic and zeolitic soils using 1 N NaOAc (pH 7) for saturating the soils, and 1 N NH_4OAc (pH 7) for exchanging the Na^+ ions; and the CEC was determined by estimating the adsorbed Na^+ ions (Richards 1954). But for the calcareous, and slight to moderately alkaline Vertisols, determination of extractable Ca and Mg is done following 1 N NaCl solution extraction method (Piper 1966) and for Na and K, 1 N NH_4OAc (pH 7) (Pal et al. 2006) is used. The base saturation (BS) calculated using CEC and extractable bases exceeds >100 either throughout the pedon depth or in the sub-soils (Table 6.1), confirming the presence of zeolites in general and heulandite in particular (as confirmed by the XRD technique, Bhattacharyya et al. 1993, 1999). The BS in acidic Vertisols in excess of 50, and 100 for calcareous and slightly to moderately alkaline Vertisols (Table 6.2), and ESP in excess of 100 of the IGP sodic soils provides an insight into the chemical environment of zeolitic soils even in absence of XRD and SEM facilities (Pal et al. 2013; Bhattacharyya et al. 2015)

Table 6.1 Selected chemical properties of zeolitic and acidic Alfisols of humid tropical climate

Horizon	Depth (cm)	pH(1:2)	Extractable bases cmol(+) kg^{-1}					Base saturation (%)
			Ca	Mg	Na	K	Sum	CEC	
A1	0–14	5.5	7.0	3.0	0.4	0.7	11.1	17.0	65
B21t	14–40	5.2	7.0	2.5	0.4	0.4	10.3	17.0	61
B22t	40–97	5.5	7.0	3.0	0.4	0.4	10.8	16.0	67
B23t	97–151	5.7	6.5	3.5	0.4	0.3	10.7	15.0	71
C	151+	5.7	8.0	4.0	0.5	0.5	13.0	12.0	108

(Adapted from Bhattacharyya et al. 1993)

Table 6.2 Selected chemical properties of zeolitic Sodic Haplusterts of SAT climate

Horizon	Depth(cm)	pH (1:2)	Extractable bases cmol(+) kg^{-1}					Base saturation (%)
			Ca	Mg	Na	K	Sum	CEC	
Ap	0–11	8.2	21.1	9.8	1.0	0.7	32.6	27.6	118
Bw1	11–37	8.4	20.4	8.9	1.2	0.6	31.1	27.5	113
Bw2	37–63	8.7	18.0	13.1	2.6	0.5	34.2	28.5	120
Bss1	63–98	8.8	14.4	13.8	4.7	0.5	33.4	29.0	115
Bss2	98–145	8.6	12.7	15.6	8.5	0.5	37.3	30.3	123
BC	145–160	8.5	11.8	14.0	10.1	0.5	36.4	32.3	112

(Adapted from Pal et al. 2009a)

6.5 Palygorskite and its Role in Soil Classification and Management

In general, SAT Vertisols have drainage problem as their sHC decreases down the depth; however the sHC decreases sharply in soils with subsoil sodicity (ESP > 5) (See Chap. 4, Fig. 4.1). In non-sodic Vertisols, the observed poor drainage is due to clay dispersion caused by exchangeable magnesium (Vaidya and Pal 2002), which suggest that the saturation of Vertisols not only with Na$^+$ ions but also with Mg^{2+} ions blocks small pores in the soil (Pal et al. 2006). In other words, Mg^{2+} ions are less efficient than Ca^{2+} ions at flocculating soil colloids, although the United States Salinity Laboratory (Richards 1954) grouped Ca^{2+} and Mg^{2+} ions together as both these bi-valent cations are expected to improve the soil structure. The blocking of small pores is further impaired by low ESP (>5, <15) which reduces the sHC to <5 mm h^{-1}, causing more than 50% reduction in cotton yield (Pal et al. 2012a).The dispersion of clay colloids and impairment of the sHC of Vertisols is generally an effect of ESP or EMP (exchangeable magnesium percentage) in presence or absence of soil modifiers. Interestingly, the sHC of zeolitic Vertisols of the Marathwada region of Maharashtra is reduced to less than 10 mm h^{-1}, although the soils are non-sodic (Typic Haplusterts, Zade 2007), neutral to mildly alkaline in pH, with less than 5 ESP, and EMP increasing with depth (Table 6.3). In some pedons, the EMP is greater than ECP (exchangeable calcium percentage) at depths below 50 cm because of the presence of silt and coarse clay palygorskite (Zade 2007). Palygorskite

6.5 Palygorskite and its Role in Soil Classification and Management

Table 6.3 Selected chemical properties of palygorskite containing Typic Haplusterts of SAT Maharashtra

Horizon	Depth (cm)	Exch. Ca/Mg	ECP[1]	EMP[2]	ESP[3]	WDC[4]%	Exch. Ca/(Mg + Na)	Base saturation %
Ap	0–10	4.9	80.8	16.3	0.5	20.7	4.8	114
Bw1	10–34	4.1	78.5	19.3	0.3	22.4	4.0	112
Bw2	34–66	3.7	77.3	20.6	0.5	22.6	3.7	115
Bss1	66–89	2.7	71.4	26.0	0.5	20.4	2.7	113

[1]*ECP* = exchangeable calcium percentage
[2]*EMP* = exchangeable magnesium percentage
[3]*ESP* = exchangeable sodium percentage
[4]*WDC* = water dispersible clay (Adapted from Zade 2007)

Fig. 6.1 Representative photograph of plasmic fabric in cross polarized light of palygorskite containing Typic Haplusterts showing poor plasma separation (partly unistrial and partly mosaic speckled b-fabric) (Adapted from Zade 2007)

minerals are present in both normal black (Typic Haplusterts) and associated sodic black soils (Sodic Haplusterts) in India (Zade 2007; Kolhe et al. 2011) and elsewhere (Heidari et al. 2008). Among the commonly found minerals, palygorskite contains more magnesium (Singer 2002). Influence of clay mineralogy on disaggregation in some palygorskite-, smectite- and kaolinite containing soils (ESP < 5) of the Jordan and Betshe'an valley in Israel, was examined by Neaman et al. (1999). They observed that palygorskite is the most disaggregated of the clay minerals, and its fibre is not associated with or within aggregates in soils and suspensions even when the soils are saturated with Ca^{2+} ions. Thus, palygorskite particles translocate down the pedon depth preferentially over smectite and eventually clog the soil pores (Neaman and Singer 2004). Therefore, palygorskite containing Vertisols with high EMP lead to the dispersion of the clay colloids that form a 3D mesh in the soil matrix and restrict the swelling of clay smectites leading to poor plasma separation (Fig. 6.1), which is fairly comparable to that of Sodic Haplusterts (See Chap. 4, Fig. 4.4b). This interaction causes drainage problems when such soils are irrigated, presenting a predicament for crop production. In view of their poor drainage conditions and loss of productivity, non-sodic Vertisols (Typic Haplusterts) with palygorskite minerals may be considered as naturally degraded soils. Soils with such characteristics also occur in close association with non-sodic and non-palygorskitic Vertisols in many watersheds of India and therefore a new initiative by pedologists

6.5.1 Simple Method to Identify the Presence of Palygorskite

The standard analytical procedure generally followed for SAT Vertisols is to determine the CEC of calcareous soils by saturating the soils using 1 N NaOAc (pH 8.2), and 1 N NH$_4$OAc (pH 7) for exchanging the Na$^+$ ions; and the CEC is determined by estimating the adsorbed Na$^+$ ions (Richards 1954). Extractable Ca and Mg is done following 1 N NaCl solution extraction method (Piper 1966) and for Na and K, 1 N NH$_4$OAc (pH 7) is used. After following these procedures, it was observed that the EMP of palygorskitic but non-sodic Vertisols (Typic Vertisols), in general, increased with pedon depth while the ECP showed a decrease, which results in lowering both the exchangeable Ca/Mg and Ca/(Na + Mg) ratio (Table 6.3). On many occasions such ratios show <1 beyond 50 cm soil depth, and often, the base saturation (BS) exceeds more than 100 in both non-zeolitic and zeolitic but palygorskite containing Typic Haplusterts and Sodic Haplusterts (Zade 2007; Kolhe et al. 2011). Even with BS > 100 such soils have sHC determined following the method of Richards (1954), reduces to << 10 mm h^{-1} even in presence and absence of Ca-zeolites (Zade 2007; Kolhe et al. 2011). A higher EMP than ECP in the subsoils along with highly reduced sHC (<< 10 mm h^{-1}) in zeolitic Typic Haplusterts, can well confirm the presence of palygorskite is SAT Vertisols even without identifying it by XRD method.

References

Abrol IP, Bhumbla DR (1978) Some comments on terminology relating to salt-affected soils. In: Proc. Dryland Saline-Seep Control, 11th Int. Congr. Soil Sci., Edmonton, Canada, June 1978, pp. 6–19 and 6–27

Bhattacharyya T, Pal DK, Deshpande SB (1993) Genesis and transformation of minerals in the formation of red (Alfisols) and black (Inceptisols and Vertisols) soils on Deccan basalt in the western Ghats, India. J Soil Sci 44:159–171

Bhattacharyya T, Pal DK, Srivastava P (1999) Role of zeolites in persistence of high altitude ferruginous Alfisols of the humid tropical Western Ghats, India. Geoderma 90:263–276

Bhattacharyya T, Pal DK, Lal S, Chandran P, Ray SK (2006) Formation and persistence of Mollisols on zeolitic Deccan basalt of humid tropical India. Geoderma 146:609–620

Bhattacharyya T, Pal DK, Mandal C, Chandran P, Ray SK, Sarkar D, Velmourougane K, Srivastava A, Sidhu GS, Singh RS, Sahoo AK, Dutta D, Nair KM, Srivastava R, Tiwary P, Nagar AP, Nimkhedkar SS (2013) Soils of India: historical perspective, classification and recent advances. Curr Sci 104:1308–1323

Bhattacharyya T, Chandran P, Pal DK, Mandal C, Mandal DK (2015) Distribution of zeolitic soils in India. Curr Sci 109:1305–1313

Dieleman PJ (1963) Reclamation of salt affected soils in Iraq. Institute of Land Reclamation and Improvement. ILRI Pub. 11, Wageningen

References

Goyal VP, Ahuja RL, Bhandari RC, Siyag RS, Sharma AP (1974) Detailed soil survey of government agricultural farm Hansi and its adjoining areas. Haryana Agricultural University, Hissar, p 83

Gupta RK, Singh CP, Abrol IP (1985) Determination of cation exchange capacity and exchangeable sodium in alkali soils. Soil Sci 139:326–332

Heidari A, Mahmoodi S, Roozitalab MH, Mermut AR (2008) Diversity of clay minerals in the Vertisols of three different climatic regions in western Iran. J Agric Sci Technol 10:269–284

Kadu PR, Vaidya PH, Balpande SS, Satyavathi PLA, Pal DK (2003) Use of hydraulic conductivity to evaluate the suitability of Vertisols for deep rooted crops in semi-arid parts of Central India. Soil Use Manag 19:208–216

Kapoor BS, Singh HB, Goswami SC (1981) Analcime in sodic profile. J Indian Soc Soil Sci 28:513–515

Kolhe AH, Chandran P, Ray SK, Bhattacharyya T, Pal DK, Sarkar D (2011) Genesis of associated red and black shrink-swell soils of Maharashtra. Clay Res 30:1–11

Leffelaar PA, Sharma R (1977) Leaching of a highly saline-sodic soil. J Hydrol 32:203–218

Ming DW, Allen ER (2001) Use of natural zeolites in agronomy, horticulture and environmental soil remediation. Rev Miner Geochem 45:619–654

Ming DW, Dixon JB (1987) Quantitative determination of clinoptilolite in soils by a cation-exchange capacity method. Clay Clay Miner 35:463–468

Neaman A, Singer A (2004) The effects of palygorskite on chemical and physico-chemical properties of soils: a review. Geoderma 123:297–303

Neaman A, Singer A, Stahr K (1999) Clay mineralogy as affecting disaggregation in some palygorskite-containing soils of the Jordan and Bet-She'an valleys. Aust J Soil Res 37:913–928

Pal DK (2013) Soil modifiers: their advantages and challenges. Clay Res 32:91–101

Pal DK, Dasog GS, Vadivelu S, Ahuja RL, Bhattacharyya T (2000) Secondary calcium carbonate in soils of arid and semi-arid regions of India. In: Lal R, Kimble JM, Eswaran H (eds) Stewart BA(eds) global climate change and pedogenic carbonates. Lewis Publishers, Boca Raton, pp 149–185

Pal DK, Bhattacharyya T, Ray SK, Chandran P, Srivastava P, Durge SL, Bhuse SR (2006) Significance of soil modifiers (ca-zeolites and gypsum) in naturally degraded Vertisols of the peninsular India in redefining the sodic soils. Geoderma 136:210–228

Pal DK, Bhattacharyya T, Chandran P, Ray SK, Satyavathi PLA, Durge SL, Raja P, Maurya UK (2009a) Vertisols (cracking clay soils) in a climosequence of peninsular India : evidence for Holocene climate changes. Quatern Int 209:6–21

Pal DK, Bhattacharyya T, Chandran P, Ray SK (2009b) Tectonics-climate-linked natural soil degradation and its impact in rainfed agriculture: Indian experience. In: Wani SP, Rockström J, Oweis T (eds) Rainfed agriculture: unlocking the potential CABI international. Oxfordshire, U.K., pp 54–72

Pal DK, Tarafdar JC, Sahoo AK (2009c) Analysis of soils for soil survey and mapping. In: Bhattacharyya T, Sarkar D, Pal DK (eds) Soil survey manual. NBSS&LUP Publication No. 146, India, p 400

Pal DK, Lal S, Bhattacharyya T, Chandran P, Ray SK, Satyavathi PLA, Raja P, Maurya UK, Durge SL, Kamble GK (2010) Pedogenic thresholds in benchmark soils under rice-wheat cropping system in a climosequence of the indo-Gangetic Alluvial Plains. Final project report, division of soil resource studies. NBSS&LUP, Nagpur, p 193

Pal DK, Wani SP, Sahrawat KL (2012a) Vertisols of tropical Indian environments: pedology and edaphology. Geoderma 189-190:28–49

Pal DK, Bhattacharyya T, Wani SP (2012b) Formation and management of cracking clay soils (Vertisols) to enhance crop productivity: Indian experience. In: Lal R, Stewart BA (eds) . World soil resources, Francis and Taylor, pp 317–343

Pal DK, Wani SP, Sahrawat KL (2013) Zeolitic soils of the Deccan basalt areas in India: their pedology and edaphology. Curr Sci 105:309–318

Pal DK, Wani SP, Sahrawat KL, Srivastava P (2014) Red ferruginous soils of tropical Indian environments: a review of the pedogenic processes and its implications for edaphology. Catena 121:260–278. https://doi.org/10.1016/j.catena2014.05.02

Pal DK, Bhattacharyya T, Sahrawat KL, Wani SP (2016) Natural chemical degradation of soils in the Indian semi-arid tropics and remedial measures. Curr Sci 110:1675–1682

Piper CS (1966) Soil and plant analysis. Hans Publishers, Bombay

Richards LA (ed) (1954) Diagnosis and improvement of saline and alkali soils, USDA agriculture handbook 60, USDA, Washington, USA

Sharma DR, Mondal RC (1981) Case study on sodic hazard of irrigation waters. J Indian Soc Soil Sci 29:270–273

Shields LG, Meyer MW (1964) Carbonate clay: measurement and relationship to clay distribution and cation exchange capacity. Soil Sci Soc Amer Proc 28:416–419

Singer A (2002) Palygorskite and sepiolite. In: Dixon JB, Schulze DG (eds) Soil mineralogy with environmental applications, SSSA Book Series, vol 7. Soil Science Society of America, Madison, pp 555–583

Srivastava P, Bhattacharyya T, Pal DK (2002) Significance of the formation of calcium carbonate minerals in the pedogenesis and management of cracking clay soils (Vertisols) of India. Clay Clay Miner 50:111–126

Srivastava P, Pal DK, Aruche KM, Wani SP, Sahrawat KL (2015) Soils of the indo-Gangetic Plains: a pedogenic response to landscape stability, climatic variability and anthropogenic activity during the Holocene. Earth-Sci Rev 140:54–71. https://doi.org/10.1016/j.earscirev.2014.10.010

Srivastava P, Aruche M, Arya A, Pal DK, Singh LP (2016) A micromorphological record of contemporary and relict pedogenic processes in soils of the indo-Gangetic Plains: implications for mineral weathering, provenance and climatic changes. Earth Surf Proc Land 41:771–790. https://doi.org/10.1002/esp.3862

Vaidya PH, Pal DK (2002) Microtopography as a factor in the degradation of Vertisols in Central India. Land Degrad Dev 13:429–445

Yaalon DH (1958) Studies on the effect of saline irrigation water on calcareous soils. II The behavior of calcium carbonate Bull Council Irs 7G:115–122

Zade SP (2007) Pedogenic studies of some deep shrink-swell soils of Marathwada region of Maharashtra to develop a viable land use plan, Ph. D thesis, Dr. PDKV, Akola, Maharashtra

Chapter 7
Mineralogy Class of Indian Tropical Soils

Abstract It is well understood that a thorough knowledge and appreciation of minerals in soils is critical for their proper use and management. Proper understanding on the role of minerals in soils has become almost mandatory to investigate the properties of the minerals, especially clay minerals, their mixtures and surface modifications in the form that they occur in the soil. While assigning the mineralogy class of soils, it is often observed that some of the important physical and chemical properties such as COLE for soils on basic igneous rocks, cation exchange capacity (CEC) of acid soils derived from soil CEC determined by $BaCl_2$-TEA (pH 8.2) plus CEC by NH_4OAc (pH 7) and effective CEC (ECEC) derived from sum of bases by NH4OAc (pH 7) plus IN KCl extractable Al^{+3}, do not get due recognition. Although with the use of high resolution mineralogical tool like XRD both identification of minerals and some enigmatic soil mineralogical classes can be conveniently solved, such instrumental facilities are generally rare in national soils' laboratory. Therefore, in this chapter some simple analytical methods are showcased, which are also useful in assigning proper mineralogy class.

Keywords Indian tropical soils · Mineralogy class of soils · Use of simple methods

For a long time many continued to believe that soils of the humid tropical (HT) climate as exemplified by highly deep red and highly weathered soils (Eswaran et al. 1992) are not conducive for plant growth and productivity (Aleva 1994) because of their high acidity and kaolinitic/gibbsitic mineralogy (Schwertmann and Herbillon 1992). In addition, crops suffer from Fe and Al toxicity (Sehgal 1998). Kaolinitic mineralogy class is assigned to many Alfisols and Ultisols of Kerala, Karnataka and Tamil Nadu of tropical HT climate of India (Bhattacharyya et al. 2009) based on their clay CEC (by 1 N NH_4OAc, pH 7) and ECEC (sum of bases extracted with 1 N NH_4OAc, pH 7 plus 1 N KCl extractable Al^{+3}) values, which are less than 16 and 12 cmol (p^+) kg^{-1}, respectively (Smith 1986), and thus they are LAC (low activity clay) soils, supporting the existence of Kandic horizon. Recent research using high resolution X-ray diffraction (XRD) and scanning electron microscopy (SEM) on soil particle size fractions of the acidic bench mark red

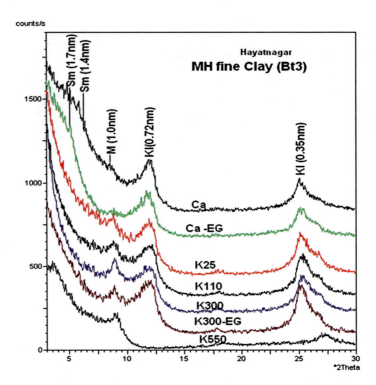

Fig. 7.1 Representative XRD of fine clay (<0.2 μm) of Typic Rhodustalfs southern India Ca- Ca saturated, Ca-EG: Ca saturated and glycolated, K25, K110, K300, K 550, K300EG: K-saturated and heated to 25,110,300, 550 °C and glycolated respectively. Sm = smectite, Kl = kaolin, M = mica (Adapted from Chandran et al. 2013)

ferruginous (RF) soils (Ultisols, Alfisols and Inceptisols) of the HT climate of India (Pal et al. 2014) in the states of Maharashtra (Bhattacharyya et al. 1993, 1999), Madhya Pradesh (Bhattacharyya et al. 2005, 2006a), Karnataka (Kharche 1996), Kerala (including soils very close to Angadipuram-the type locality of laterite, a name first coined by Francis Buchanan in 1800, Chandran et al. 2005a), Goa (Chandran et al. 2004), Jharkhand (Ray et al. 2001), Meghalaya (Bhattacharyya et al. 2000), Tripura (Bhattacharyya et al. 2006b), Manipur (Chandran et al. 2006), Assam (Pal et al. 1987) and Andaman and Nicobar islands (Chandran et al. 2005b) indicates the dominant presence of kaolin (Kl-HIV/HIS) (a 0.7 nm mineral interstratified with hydroxy-interlayered vermiculite, HIV or smectite, HIS) (Fig. 7.1) and not a true kaolinite, with occasional presence of gibbsite (Pal et al. 2014). The 0.72 nm peak of kaolin has a broad base, tailing towards the low angle region. On heating the K-saturated sample at 550 °C, the 0.72 nm peak disappeared, confirming the presence of kaolin and simultaneously reinforced the 1.0 nm region. In view of the very small amount of 1.4 nm minerals, the higher degree of reinforcement of the 1.0 nm region at 550 °C compared with that at 300 °C may be possible only when the 0.7 nm minerals are Kl-HIV/HIS. The presence of kaolin is in contrast to the general perception that these soils are dominated by kaolinite and/or gibbsite. These soils also contain considerable amounts of weatherable minerals (>10%) like mica, mica-hydroxy-interlayered smectite (M-HIS), and hydroxy-interlayered vermiculite/smectite (HIV/HIS) (Fig. 7.1) (Chandran et al. 2004, 2005a, 2013). The prevailing acid weathering causes hydroxy-interlayering of vermiculite/smectites of these soils. Because of hydroxy-interlayering, the acidity of soils determined by $BaCl_2$-TEA is much higher than that determined by using 1 N KCl. For many such soils, this total acidity ($BaCl_2$-TEA) plus the sum of bases by NH_4OAc (pH 7) (clay CEC of sum of cations in SCS, soil control section) indicates a value of ≥24, a value much greater than 12 (Table 7.1) (Chandran et al. 2005a; Pal et al. 2014). Based on semi-quantitative estimates of clay gibbsite in SCS of Ultisols of Kerala, a gibbsitic /allitic mineralogy class could be suggested following the US Soil Taxonomy. However, such mineralogy class would undermine the contemporary pedogenesis of the formation of hydroxy-interlayered clay minerals. Therefore, the most appropriate mineralogy class for Ultisols (containing considerable amount of gibbsite) of the HT climate should be 'mixed' as this class is compatible with their present use for horticultural, forestry and agricultural crops.

Some of acidic Alfisols developed on Deccan basalts of Ratnagiri district of the Konkan regions of Maharashtra state have clay CEC of 21 cmol (p^+) kg^{-1} and according to this value kaolinitic mineralogy class (Smith 1986) is assigned to these soils (Bhattacharyya et al. 2009; Lal 2000). Total acidity of such soils by $BaCl_2$-TEA (Peech et al. 1947; Jackson 1973) was not determined (Lal 2000). However, it is interesting to note that such soils have >50% base saturation and also have COLE > > 0.06 because of the considerable amount of hydroxy-interlayered smectite (HIS) in kaolin dominated soils (a 0.7 nm mineral interstratified with HIS). It is obvious that clay CEC value is grossly underestimated. Addition of total acidity by $BaCl_2$-TEA to the extractable bases by 1 N NH_4OAc, the clay CEC value would have been much greater than 21 cmol (p^+) kg^{-1} to suggest the smectitic mineralogy class (Smith 1986). It is now a well-established fact that for soil with a COLE >0.06 must have at least 20% clay smectite (Fig. 7.2), which is enough to cause vertic properties in soils (Shirsath et al. 2000). Large amount of Al_3^+ ions released during HT weathering is trapped in clay minerals other than kaolinite, which is not easily extractable by 1 N unbuffered KCl solution (Pal et al. 2014). Thus, this situation demands the determination of total acidity by $BaCl_2$-TEA and its addition to extractable bases by 1 N NH_4OAc to derive the true clay CEC of soils of HT climate and therefore, the use of clay CEC (Smith 1986) for assigning the mineralogy class for kaolin (Kl-HIV/HIS) dominated soils of humid tropical Western Ghats is not justified. Smectitic mineralogy class of acidic Alfisols on Deccan basalt can fully justify their potential use for the present horticultural, forestry and agricultural land uses like the HT soils of other Indian states. What follows from the above is to assign mineralogy class for kaolin dominated HT soils, acquisition of total acidity data by $BaCl_2$-TEA method (Peech et al. 1947; Jackson 1973) alongside extractable bases by 1 N NH_4OAc (Richards 1954), is just adequate enough to address the real mineralogy class, which would also explain satisfactorily their present land uses.

Table 7.1 Physical and chemical properties of selected RF soils of HT climate

Horizon	Depth cm	pH H$_2$O	pH KCl	ΔpH	Clay %	Organic carbon %	Exchange acidity KCl(N) H$^+$ cmol(P$^+$)/kg	Exchange acidity KCl(N) Al^{3+} cmol(P$^+$)/kg	Extractable acidity, BaCl$_2$-TEA	CEC, soil (NH$_4$OAC,7)	ECEC, soil (NH$_4$OAC,7)	CEC, clay	ECEC, clay	Base saturation, %
Ustic Kandihumults: Kerala[a]
Ap	0–13	4.8	4.3	−0.5	21.1	2.35	0.33	0.60	11.2	4.5	1.6	21.3	7.6	22
Bt1	13–32	4.4	4.3	−0.1	31.3	1.86	0.50	0.48	10.4	3.5	1.1	11.2	3.5	17
Bt2	32–56	4.5	4.3	−0.2	29.0	1.50	0.43	0.32	10.0	3.7	0.8	12.8	2.7	14
Bt3	56–83	4.5	4.6	+0.1	26.0	0.90	0.23	0.12	9.0	4.1	0.7	16.0	2.7	13
Bt4	83–112	4.4	4.6	+0.2	28.5	1.11	0.20	0.10	6.6	4.0	0.6	13.7	2.1	13
Bt5	112–150	4.7	4.7	Nil	24.0	1.22	0.30	0.10	7.0	4.0	0.7	16.7	3.0	17
Kanhaplohumults: Arunachal Pradesh[b]
A1	0–13	4.6	3.8	−0.8	53.0	2.70	2.0	0.7	23.3	13.2	4.8	23.0	9.2	16
Bt1	13–36	4.8	3.8	−0.7	66.0	1.96	2.2	1.0	18.5	12.3	4.1	18.6	6.2	18
Bt21	36–63	5.0	4.5	−0.8	68.0	0.87	2.4	1.0	14.2	10.5	4.0	15.4	6.0	16
Bt22	63–100	5.4	4.2	−0.8	65.0	0.64	2.6	1.1	14.3	9.7	4.4	15.0	6.8	17
B3	100–125	5.2	4.2	−1.0	47.5	0.29	2.7	1.1	10.2	7.1	4.4	15.0	9.3	18
Kanhaplohumults: Assam[b]
A1	0–14	5.0	3.8	−1.2	48.5	2.20	1.4	0.4	16.1	11.0	5.6	22.4	11.5	34
Bt11	14–33	4.8	3.7	−1.1	56.5	1.60	2.2	0.9	16.0	10.5	5.6	18.6	10.0	24
Bt12	33–85	5.1	3.8	−1.3	62.0	1.00	2.0	0.9	13.0	9.0	4.4	14.3	7.1	17
Bt21	85–120	5.2	4.0	−1.2	65.5	0.70	1.5	0.6	12.2	7.7	4.0	11.7	6.0	21

7 Mineralogy Class of Indian Tropical Soils

BC	120–180	5.4	4.0	−1.4	63.5	0.50	1.3	0.5	12.0	6.5	3.1	10.2	5.0	19
Typic Dystrochrepts: Assam[b]														
A1	0–13	5.0	3.9	−1.1	29.5	1.50	1.4	0.3	10.7	10.0	5.4	32.5	18.2	37
A2	13–31	4.8	3.9	−0.9	34.5	0.80	2.3	0.9	9.3	7.5	5.5	21.7	16.0	30
B21	31–70	4.9	3.9	−1.0	34.0	0.66	2.1	0.7	10.6	7.0	4.8	20.3	14.0	22
B22	70–122	4.7	4.1	−0.6	44.5	0.57	2.1	0.7	10.7	8.1	4.6	18.2	10.0	18
B3	122–175	5.0	4.1	−0.9	29.0	0.32	1.5	0.4	11.6	5.3	3.3	18.2	12.2	29
Typic Dystrochrepts: Manipur[b]														
A1	0–10	4.0	3.6	−0.4	35.0	1.60	2.4	1.1	10.5	9.2	7.8	26.3	22.3	51
A2	10–28	4.0	3.6	−0.4	40.0	1.20	2.9	1.4	8.7	8.1	6.3	20.2	15.7	28
B21	28–60	4.2	3.8	−0.4	40.5	0.70	3.3	1.5	8.8	8.0	5.8	19.5	14.3	26
B22	60–90	4.2	3.8	−0.4	39.0	0.60	3.1	1.5	8.6	7.6	5.5	19.5	14.0	30
BC	90–125	4.2	3.9	−0.3	28.0	0.30	2.7	1.3	6.5	7.8	5.0	27.8	17.5	13
Typic Kandihumults:Meghalaya[b]														
A1	0–16	4.5	4.2	−0.3	21.6	3.6	1.14	0.36	32.0	7.3	2.5	33.8	16.6	19
B21	16–31	4.8	4.4	−0.4	30.4	2.5	0.64	0.16	27.0	7.3	3.0	24.0	10.0	20
B22	31–62	5.0	4.8	−0.2	31.1	2.0	0.30	Nil	22.0	4.1	0.8	13.2	2.6	19
B23	62–95	5.1	6.4	+0.9	26.5	0.6	0.10	Nil	15.8	4.1	1.2	15.5	4.5	29

[a]Adapted from Chandran et al. (2005a)
[b]Adapted from Sen et al. (1997)

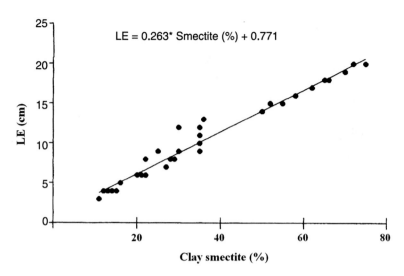

Fig. 7.2 Relationship between linear extensibility (LE) and smectite content of soils with vertic characters. (Adapted from Shirsath et al.2000)

Assigning the smectite mineralogy class for SAT Vertisols based on clay CEC and COLE data is an easy task but the same may not be straight forward for soils like shallow and medium, non-acidic SAT cracking clay soils developed on the Deccan basalts (Entisols and Vertic Inceptisols and Alfisols) that often reported without their COLE data. In view of the above relation between COLE and clay smectite content (Fig. 7.2), a value of COLE >0.06 can ensure the vertic character of soils. If soils have vertic character, the mineralogy class for such soils will be smectitic (Shirsath et al. 2000). Thus, the use of sophisticated analytical instruments for mineralogical analysis may not be always necessary.

References

Aleva GJJ (1994) Laterites: concepts, geology, morphology and chemistry. In: Creutzberg D (ed) International soil reference and information Centre (ISRIC). Wageningen, The Netherlands

Bhattacharyya T, Pal DK, Deshpande SB (1993) Genesis and transformation of minerals in the formation of red (Alfisols) and black (Inceptisols and Vertisols) soils on Deccan basalt in the western Ghats, India. J Soil Sci 44:159–171

Bhattacharyya T, Pal DK, Srivastava P (1999) Role of zeolites in persistence of high altitude ferruginous Alfisols of the western Ghats, India. Geoderma 90:263–276

Bhattacharyya T, Pal DK, Srivastava P (2000) Formation of gibbsite in presence of 2:1 minerals: an example from Ultisols of Northeast India. Clay Miner 35:827–840

Bhattacharyya T, Pal DK, Chandran P, Ray SK (2005) Land-use, clay mineral type and organic carbon content in two Mollisols–Alfisols–Vertisols catenary sequences of tropical India. Clay Res 24:105–122

Bhattacharyya T, Pal DK, Lal S, Chandran P, Ray SK (2006a) Formation and persistence of Mollisols on Zeolitic Deccan basalt of humid tropical India. Geoderma 136:609–620

References

Bhattacharyya T, Pal DK, Velayutham M, Vaidya P (2006b) Sequestration of aluminium by vermiculites in LAC soils of Tripura. Abstract, 71st annual convention and National Seminar on "developments of soil science" of the Indian Society of Soil Science, Bhubaneswar, Orissa, p 1

Bhattacharyya T, Sarkar D, Sehgal JL, Velayutham M, Gajbhiye KS, Nagar AP, Nimkhedkar SS (2009) Soil Taxonomic Database of India and the States (1:250,000 scale), NBSSLUP, Publication 143, p 266

Chandran P, Ray SK, Bhattacharyya T, Dubey PN, Pal DK, Krishnan P (2004) Chemical and mineralogical characteristics of ferruginous soils of Goa. Clay Res 23:51–64

Chandran P, Ray SK, Bhattacharyya T, Srivastava P, Krishnan P, Pal DK (2005a) Lateritic soils of Kerala, India: their mineralogy, genesis and taxonomy. Aust J Soil Res 43:839–852

Chandran P, Ray SK, Bhattacharyya T, Pal DK (2005b) Chemical and mineralogical properties of ferruginous soils of Andaman and Nicobar Islands. Abstract, 70th annual convention and National Seminar on"developments of soil science" of the Indian Society of Soil Science. TNAU, Coimbatore, Tamil Nadu, p 45

Chandran P, Ray SK, Bhattacharyya T, Sen TK, Sarkar D, Pal DK (2006) Rationale for mineralogy class of ferruginous soils of India. Abstract, 15th annual convention and National Symposium on "clay research in relation to agriculture, environment and forestry" of the clay minerals Society of India. BCKVV, Mohanpur, West Bengal, p 1

Chandran P, Ray SK, Bhattacharyya T, Tiwari P, Sarkar D, Pal DK, Mandal C, Nimkar A, Maurya UK, Anantwar SG, Karthikeyan K, Dongare VT (2013) Calcareousness and subsoil sodicity in ferruginous Alfisols of southern India: an evidence of climate shift. Clay Res 32:114–126

Eswaran H, Kimble J, Cook T, Beinroth FH (1992) Soil diversity in the tropics: implications for agricultural development. In: Lal R, Sanchez PA (eds), Myths and science of soils of the tropics. SSSA Special Publication Number 29. SSSA, Inc. and ACA, Inc., Madison, pp 1–16

Jackson ML (1973) Soil chemical analysis. Prentice Hall of India Pvt Ltd, New Delhi

Kharche VK (1996) Developing soil-site suitability criteria for some tropical plantation crops. Ph. D thesis. Dr. P D K V., Akola

Lal S (2000) Characteristics, genesis and use potential of soils of the western Ghats, Maharashtra. Ph. D thesis. Dr. P D K V. Akola

Pal DK, Deshpande SB, Durge SL (1987) Weathering of biotite in some alluvial soils of different agro climatic zones. Clay Res 6:69–75

Pal DK, Wani SP, Sahrawat KL, Srivastava P (2014) Red ferruginous soils of tropical Indian environments: a review of the pedogenic processes and its implications for edaphology. Catena 121:260–278. https://doi.org/10.1016/j.catena2014.05.023

Peech M, Alexander LT, Dean LA, Reed JF (1947) Methods of soil analysis and soil fertility investigations. U S Department of Agriculture, Circular No.752

Ray SK, Chandran P, Durge SL (2001) Soil taxonomic rationale: kaolinitic and mixed mineralogy classes of highly weathered ferruginous soils. Abstract, 66th Annual Convention and National Seminar on "Developments in soil science" of the Indian Society of Soil Science, Udaipur, Rajasthan, pp 243–244

Richards LA (ed) (1954) Diagnosis and improvement of saline and alkali soils, USDA Agricultural Handbook, vol 60. US Government Printing Office, Washington, DC

Schwertmann U, Herbillon AJ (1992) Some aspects of fertility associated with the mineralogy of highly weathered tropical soils. In: Lal R, Sanchez PA (eds) Myths and science of soils of the tropics, SSSA Special Publication Number SSSA 29. Inc and ACA, Inc, Madison, pp 47–59

Sehgal JL (1998) Red and lateritic soils: an overview. In: Sehgal J, Blum WE, Gajbhiye KS (eds) Red and lateritic soils. Managing red and lateritic soils for sustainable agriculture, vol 1. Oxford and IBH Publishing Co. Pvt. Ltd., New Delhi, pp 3–10

Sen TK, Nayak DC, Dubey PN, Chamuah GS, Sehgal JL (1997) Chenical and electrochemical characterization of some acid soils of Assam. J Indian Soil Sci 45:245–249

Shirsath SK, Bhattacharyya T, Pal DK (2000) Minimum threshold value of smectite for vertic properties. Aust J Soil Res 38:189–201

Smith GD (1986) The Guy Smith Interviews: Rationale for concept in soil taxonomy. SMSS Technical Monograph, 11. SMSS, SCS, USDA, USA

Chapter 8
Hydraulic Conductivity to Evaluate the SAT Vertisols for Deep-Rooted Crops

Abstract Recent attempts to identify bio-physical factors that limit the yield of deep-rooted crops (cotton) in non-gypsic and non-zeolitic SAT Vertisols in the Vidarbha region in central India indicate that the retention and release of soil water are governed by the nature and content of clay minerals, and also by the nature of the exchangeable cations. The available water content (AWC) is not available during the growth of crops. In fact, moisture remains at 100 kPa for non-sodic Vertisols with ESP < 5 after the cessation of rains while it is held at 300 kPa for sodic Vertisols with ESP > 5 but <15. This unfavourable AWC is created because the movement of water is governed by saturated hydraulic conductivity (sHC), which decreases rapidly with depth, and the decrease is sharper in sodic Vertisols even with ESP > 5. Drastic reduction in sHC in the subsoils is due to the formation of pedogenic calcium carbonate (PC) with the concomitant increase of ESP with pedon depth. A significant positive correlation between the yield of cotton and carbonate clay indicates that, like ESP, PC formation also causes a yield reduction. The strong pedogenic relationship does exist among SAT environments, PC formation, exchangeable Ca/Mg, ESP and sHC, which ultimately make SAT Vertisols poorly drained. Under rain-fed conditions, the yield of deep-rooted crops on Vertisols depends primarily on the amount of rain stored in the subsoils and the extent to which this soil water is released between the rains during crop growth. In view of this, the evaluation of SAT Vertisols for deep-rooted crops based on sHC alone may help in their planning and management in the Indian SAT areas and also under similar climatic conditions elsewhere.

Keywords SAT Vertisols · Saturated hydraulic conductivity · Evaluation for deep rooted crops

Vertisols represent a large crop production resource in many countries and are generally productive, but difficult to manage. They are important soils, yielding high-to-moderate agricultural production in Madhya Pradesh, Maharashtra, Andhra Pradesh and northern Karnataka. Cereal grain yields of 3 t ha^{-1} are not uncommon under rainfed conditions. These soils are used for many purposes, including the production of cotton, sorghum and citrus. In SAT Vertisols subsoil porosity and

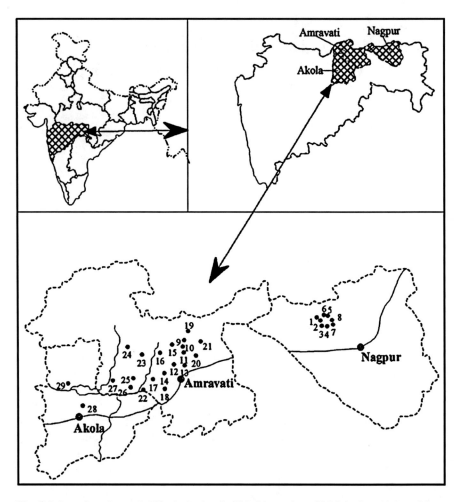

Fig. 8.1 Location of sampled Vertisols sites in Vidarbha region of Maharashtra (Adapted from Kadu et al.2003)

aeration are generally poor and roots of annual crops do not penetrate deeply. Therefore they are often difficult to cultivate, particularly for small farmers using handheld or animal-drawn implements. Farmers faced with these difficulties allow these soils to lie fallow for one or more rainy seasons or cultivate them only in the post-rainy season. Thus Vertisols have limitations that restrict their full potential to grow both rainy season and winter crops (NBSS&LUP-ICRISAT 1991) as is also reported from Vidarbha region of especially in the districts of Nagpur, Amravati and Akola of Maharashtra state of central India (Fig. 8.1). Either rainy season or winter crops are grown in Vertisols of the western part of the Amravati district and the adjoining Akola district, whereas they are grown in those of Nagpur district with limited irrigation (NBSS&LUP-ICRISAT 1991). These agricultural land uses clearly highlight that even though Vertisols make up a relatively homogeneous major soil group; they show a considerable variability in their land use and crop productivity. It is therefore important to understand the factors that cause the variability in their properties.

Attempts to identify the bio-physical factors that limit the yield of deep-rooted crops including cotton in Vertisols were made in the past (Sehgal 1991; NBSS&LUPICRISAT 1991; NBSS&LUP 1994; Mandal et al. 2002). However, some of the factors chosen were qualitative (NBSS&LUP-ICRISAT 1991; Mandal et al. 2002) and some of them (Sehgal 1991; NBSS&LUP 1994) need revision. A synthesis of recent developments in the pedology of Vertisols achieved through the use of high resolution micro-morphology, mineralogy, and age control data along with their geomorphologic and climatic history, has helped better understand the effects of pedogenetic processes due to changes in climate during the Holocene in modifying the soil properties in the presence or absence of soil modifiers (Ca-zeolites and gypsum), $CaCO_3$ and palygorskite minerals. In view of recent information on the impairment of hydraulic properties of Vertisols due to the formation of pedogenic calcium carbonate (PC) and the concomitant development of subsoil sodicity (ESP >5, <15) (Balpande et al. 1996; Pal et al. 2001; Vaidya and Pal 2002) a new initiative is required to identify the modified soil properties of Vertisols that may be included to explain the performance of deep-rooted crops under rainfed conditions.

Recently, Kadu et al. (2003) and Deshmukh et al. (2014) attempted to identify bio-physical factors that limit the yield of deep-rooted crops (cotton) in 32 Vertisols (developed in the basaltic-alluvium) of the Nagpur, Amravati and Akola districts in the Vidarbha region in central India. Judging by their pH, ECe and ESP values, many Vertisols pedons of Amravati and Akola qualify as sodic (Richards 1954) and are Sodic Haplusterts (Soil Survey Staff 1999). About 8 Vertisols of Nagpur are Typic Haplusterts/Calciusterts whereas some Vertisols of Amravati and Akola is Aridic Haplusterts. This indicates that under the ustic soil moisture regime and hyperthermic temperature regime, the soils of adjoining districts have quite different chemical environments even where they are not affected by a seasonal water table. Under similar soil management by farmers, and also under similar soil moisture and temperature regimes, yields of cotton were better in soils of Nagpur than in those of Amravati and Akola (please refer to Table 3, Kadu et al. 2003). This result is, however, not observed when soils are evaluated for cotton following the criteria of landscape and soil requirements of Sys et al. (1993). Due to low content of organic carbon (<0.8%) and imperfect drainage, not only the Sodic Haplusterts, but also the Typic and Aridic Haplusterts qualify for S3 and N1 categories (please refer to Table 3, Kadu et al. 2003). This is, however, difficult to reconcile with the yield of cotton on Typic and Aridic Haplusterts (please refer to Table 3, Kadu et al. 2003). NBSS&LUP (1994) modified some of Sys's criteria including waterlogging, available water content (AWC) and a new limit of ESP. The soils were evaluated following this modified method (NBSS&LUP 1994), and it was observed that both Typic Haplusterts and Aridic Haplusterts qualify for the S2 category, even though the yield varies from the 18 to 6 q ha^{-1} (please refer to Table 3, Kadu et al. 2003). When

Table 8.1 Co-efficient of correlation among various soil attributes and yield of cotton[a]

No.	Parameter Y	Parameter X	r
Based on 165 soil horizons samples of 29 Vertisols			
1	sHC (mm h^{-1})	Exch. Ca/Mg	0.51*
2	sHC (mm h^{-1})	ESP [b]	−0.56*
3	ESP	AWC (%)	0.40*
4	ESP	Exch. Ca/Mg	−0.40*
Based on 29 Vertisols			
5	Yield of cotton (q ha^{-1})	AWC (%) WM [c]	−0.10
6	Yield of cotton (q ha^{-1})	ESP max [a]	- 0.74*
7	Yield of cotton (q ha^{-1})	sHC WM [b]	0.76*
8	Yield of cotton (q ha^{-1})	carbonate clay [d]	−0.64*
10	Yield of cotton (q ha^{-1})	Exch. Ca/Mg WM [c]	0.50*
11	ESP max [a]	AWC (%) WM [c]	0.30*
12	ESP max [a]	Exch. Ca/Mg WM [c]	−0.55*
13	ESP max[a]	carbonate clay [d]	0.83*

[a]Adapted from Kadu et al. (2003)
[b]Maximum in pedon
[c]weighted mean
[d]fine earth basis
AWC available water content; *ESP* exchangeable sodium percentage; *sHC* saturated hydraulic conductivity
*Significant at 1% level

the method based on the percentage of the optim yield of crop (FAO 1983) was followed, all Typic Haplusterts qualified for the S1 or S2 categories and the majority of Aridic Haplusterts for the S2 or S3 category but few Aridic Haplusterts of Amravati in the S1 category (please refer to Table 3, Kadu et al. 2003). These observations suggest that the existing methods of evaluation cannot explain the substantial variation in yield from Aridic Haplusterts, owing to the limitation of soils as the yield reducing factor, as discussed later. Mandal et al. (2001) also observed reduction in the yield of sorghum in similar soils for the same reason.

Under rain-fed conditions, the yield of deep-rooted crops on Vertisols depends primarily on the amount of rain stored at depth in the soil profile and the extent to which this soil water is released between the rains during crop growth. Both the retention and release of soil water are governed by the nature and content of clay minerals, and also by the nature of the exchangeable cations. The AWC, calculated based on moisture content, varied between 33 and 1500 kPa (Table 8.1), indicating that not only the Typic/Aridic Haplusterts but also the Sodic Calciusterts can hold sufficient water; however, a non-significant negative correlation between cotton yield and AWC (Table 8.1) indicates that this water is not available during the growth of crops. Studies on biophysical factors on water retention and release, and cotton yield in 32 SAT Vertisols indicate that the determination of AWC and plant available water capacity (PAWC) at 33 and 1500 kPa overestimates the soil moisture content because in field conditions soil water in the subsoil do not reach the saturation at 33 kPa due to low to very low saturated hydraulic conductivity (Deshmukh et al. 2014). The Na$^+$ ions on exchange sites of Aridic Haplusterts and Sodic Haplusterts/Calciusterts with ESP >5 thus cause the over estimation of water (Gardner et al. 1984). In fact, moisture remains at 100 kPa for Typic Haplusterts and Aridic Haplusterts (ESP < 5) after the cessation of rains during June to September, while it is held at 300 kPa for Sodic Haplusterts (Pal et al. 2012; Deshmukh et al. 2014) as the favourable movement of water and the saturation of subsoils at 33 kPa are governed by saturated hydraulic conductivity (sHC). The sHC decreases rapidly with depth, and the decrease is sharper in Aridic/Sodic Haplusterts (ESP > 5, Pal et al. 2009) (See Chap. 4, Fig. 4.1).This conclusion is supported by a significant positive correlation between PAWC (estimated at 100–1500 kPa for non-sodic and 300–1500 kPa for sodic soils), and yield of cotton (Deshmukh et al. 2014), between ESP and AWC (moisture content between 33 and 1500 kPa), and a significant negative correlation between yield and ESP (Table 8.1). A significant positive correlation between yield and exchangeable Ca/Mg (Table 8.1) indicates that a dominance of Ca^{2+} ions in the exchange sites of Vertisols is required to improve the hydraulic properties for a favourable growth and final yield of crops. The development of subsoil sodicity (ESP ≥ 5) replaces Ca^{2+} ions in the exchange complex, causing a reduction in the yield of cotton in Aridic/Sodic Haplusterts (ESP ≥ 5). A significant negative correlation between ESP and exchangeable Ca/Mg (Table 8.1) indicates an impoverishment of soils with Ca^{2+} ions during sodification by the illuviation of Na-rich clays. This pedogenetic process depletes Ca^{2+} ions from the soil solution in the form of CaCO$_3$, with the concomitant increase of ESP with pedon depth. Thus, these soils contain PC (Pal et al. 2000), and carbonate clay, which, on a fine earth basis, increases with depth (See Chap. 3, Fig. 3.2b, c) (Pal et al. 2003). This chemical process is evident from the positive correlation between ESP and carbonate clay (Table 8.1). A significant positive correlation between the yield of cotton and carbonate clay (Table 8.1) indicates that, like ESP, PC formation also causes a yield reduction and is a more important yield influencing soil parameter than <2 mm soil CaCO$_3$ (NBSS&LUP 1994; Sys et al. 1993). Rapid formation of PC in dry climates impairs the hydraulic properties of Vertisols. This fact is evident from the significant negative correlation between ESP and sHC and also from a significant positive correlation between the yield of cotton and sHC (Table 8.1). The unique pedogenic relationship among SAT environments, PC formation, exchangeable Ca/Mg, and ESP, ultimately impairs the drainage (sHC) of Vertisols. It then follows that the evaluation of Vertisols for deep-rooted crops based on sHC (as determined using a constant head permeameter, Richards 1954) alone may help in agricultural land use planning of deep cracking clays soils of India and elsewhere under SAT environments (Kadu et al. 2003; Pal et al. 2012).

References

Balpande SS, Deshpande SB, Pal DK (1996) Factors and processes of soil degradation in Vertisols of the Purna valley, Maharashtra, India. Land Degrad Dev 7:313–324

Deshmukh HV, Chandran P, Pal DK, Ray SK, Bhattacharyya T, Potdar SS (2014) A pragmatic method to estimate plant available water capacity (PAWC) of rainfed cracking clay soils (Vertisols) of Maharashtra, Central India. Clay Res 33:1–14

FAO (1983) Guidelines: land evaluation for rainfed agriculture. FAO soils bulletin 52. FAO Rome 52:237

Gardner EA, Shaw RJ, Smith GD, Coughlan KG (1984) Plant available water capacity: concept, measurement, prediction. In: McGarity JW, Hoult EH, Co HB (eds) Properties, and utilization of cracking clay soils. University of New England, Armidale, pp 164–175

Kadu PR, Vaidya PH, Balpande SS, Satyavathi PLA, Pal DK (2003) Use of hydraulic conductivity to evaluate the suitability of Vertisols for deep-rooted crops in semi-arid parts of Central India. Soil Use Manage 19:208–216

Mandal DK, Mandal C, Velayutham M (2001) Development of a land quality index for sorghum in Indian semi-arid tropics (SAT). Agric Syst 70:335–350

Mandal DK, Khandare NC, Mandal C, Challa O (2002) Assessment of quantitative land evaluation methods and suitability mapping for cotton growing soils of Nagpur district. J Indian Soc Soil Sci 50:74–80

NBSS&LUP (1994) Proceedings of national meeting on soil-site suitability criteria for different crops. Feb 7–8, Nagpur India, p 20

NBSS&LUP-ICRISAT (1991) The suitabilities of Vertisols and associated soils for improved cropping systems in Central India. NBSS and LUP. Nagpur and ICRISAT, Patancheru, p 61

Pal DK, Dasog GS, Vadivelu S, Ahuja RL, Bhattacharyya T (2000) Secondary calcium carbonate in soils of arid and semi-arid regions of India. In: Lal R, Kimble JM, Eswaran H, Stewart BA (eds) Global climate change and pedogenic carbonates. Lewis Publishers, Boca Raton, pp 149–185

Pal DK, Balpande SS, Srivastava P (2001) Polygenetic Vertisols of the Purna Valley of Central India. Catena 43:231–249

Pal DK, Bhattacharyya T, Ray SK, Bhuse SR (2003) Developing a model on the formation and resilience of naturally degraded black soils of the peninsular India as a decision support system for better land use planning. NRDMS, Department of Science and Technology (Govt. of India) Project Report, NBSSLUP (ICAR), Nagpur. p 144

Pal DK, Bhattacharyya T, Chandran P, Ray SK (2009) Tectonics-climate-linked natural soil degradation and its impact in rainfed agriculture: Indian experience. In: Wani SP, Rockström J, Oweis T (eds) Rainfed agriculture: unlocking the potential. CABI International, Oxfordshire, pp 54–72

Pal DK, Wani SP, Sahrawat KL (2012) Vertisols of tropical Indian environments: pedology and edaphology. Geoderma 189–190:28–49

Richards LD (1954) Diagnosis and improvement of saline and alkali soils. USDA agriculture, handbook no 60, US Government Printing Office Washington DC, USA

Sehgal JL (1991) Soil-site suitability evaluation for cotton. Agropedology 1:49–63

Soil Survey Staff (1999) Soil taxonomy: a basic system of soil classification for making and interpreting soil surveys, Agriculture handbook, vol 436. United States Department of Agriculture, Natural Resource Conservation Service, U.S. Government Printing Office, Washington, DC

Sys C, van Ranst E, Debaeveye IrJ, Beernaert F (1993) Land evaluation, Part 3: Crop requirements. International training Centre for Post Graduate Soil Scientists, University of Ghent, Belgium

Vaidya PH, Pal DK (2002) Micro topography as a factor in the degradation of Vertisols in Central India. Land Degrad Dev 13:429–445

Chapter 9
Clay and Other Minerals in Selected Edaphological Issues

Abstract Many a time the description of minerals actually present in a soil is inadequate or incomplete, which makes the search for links between mineralogy and soil properties of agricultural importance, difficult. Nevertheless, it is very much necessary to investigate the properties of these minerals relevant to the properties of the soil in bulk even when soil minerals often differ from "type" minerals and also when soil clay minerals exist as their mixtures with their modified surfaces. Unless the mineralogical description is accurate enough for the purpose intended, it would be imprudent to look for their significance in soils. In the last few decades, Indian soil mineralogists with the help of high resolution mineralogy, identification and explanation of many enigmatic situations in soils could be solved. During this course of research endeavours, advanced information was developed, and also helped in developing simple analytical methods that would explain discretely many unresolved issues of the nutrient management of major tropical soil types of India and elsewhere.

Keywords Soil minerals · Characterisation of clay minerals · Nutrient management practices

9.1 Introduction

It is now well known that the clay is an important soil constituent controlling its properties. Significant research shows evidences that the amount of clay in a soil has a very important bearing on the genesis, characteristics, physical and chemical properties of soils. However, it would be more helpful to demonstrate what significance clay mineral *type* and other soil minerals have in soils. A search for links between mineralogy and edaphological issues is likely to be difficult because very often the description of minerals actually present in a soil is inadequate or incomplete. Further, it becomes more difficult as soil minerals often differ from "type" minerals. It is, however, more important to investigate the properties of these minerals relevant to the properties of the soil in bulk. In this endeavour, Pal et al. (2000a) demonstrated a good number of examples that indicated despite soil clay minerals being a mixture of several components, adequate description is possible based on research using high resolution instrumental facilities for mineralogical studies. Synthesis of the present dataset on the nature and characteristics of primary and secondary minerals of Indian soils has established a link between minerals and selected bulk soil properties (Pal et al. 2000a, 2012a, b, 2014; Srivastava et al. 2015). Additionally, such synthesis has provided enough hints to gather simple analytical methods that are capable to establish link between clay minerals and selected bulk soil properties. In the following some examples are highlighted. It is hoped that such simple analytical methods would help both pedologists and edaphologists to assess the health and quality of soils while developing suitable management practices to enhance and sustain their productivity in the twenty-first Century.

9.2 Carbon Sequestration in Indian Tropical Soils

The soil organic carbon (SOC) stock of Indian soils stored in the upper 30 and 150 cm depths (Bhattacharyya et al. 2000a) when compared to the stock for tropical regions and the world (Batjes 1996), shows that the share of Indian soils is not substantial (Bhattacharyya et al. 2008). This is because in India, there are very few organic matter rich soils like Histosols, Spodosols, Andosols and Gelisols, and the area under Mollisols is relatively small. Moreover, the soils of India cover only 11% of the total area of the world. Even under unfavourable environmental conditions for OC rich soils, the SOC stocks of Indian soils demonstrate enough potential to sequester organic C (Pal et al. 2015). Impoverishment in SOC in Indian soils, is largely due to less accumulation of organic C in soils of the arid and semi-arid and dry sub-humid climatic regions, which cover nearly 50% of the total geographical area (TGA) of India (Pal et al. 2000a) and remain as SOC deficient zones (< 1% OC) of the country. On the other hand, soils of HT climates show their OC > 1% (Velayutham et al. 2000). In Ultisols, Alfisols and Mollisols of HT climate, C sequestration is relatively high as indicated by their OC concentration ranges from 1.0 to 5% (Table 9.1) (Pal et al. 2014). These soils are developed under thermic, hyperthermic (Nilgiri Hills in Kerala and Tamil Nadu, Manipur, Meghalaya, Nagaland, Arunachal Pradesh, Assam, Tripura and Mizoram), isohyperthermic (Andaman and Nicobar, Kerala, Tamil Nadu, Madhya Pradesh and Maharashtra) soil temperature regime, and udic (Andaman and Nicobar, Arunachal Pradesh, Manipur, Meghalaya, Assam, Tripura, Nagaland, Mizoram, Nilgiri Hills in Kerala and Tamil Nadu) and ustic (Kerala, Karnataka, Tamil Nadu, Madhya Pradesh and Maharashtra) soil moisture regime (Bhattacharyya et al. 2009).

Soils in a wet climate under forest have high OC content, sufficient to qualify as Mollisols. The OC addition to Ultisols and acidic Alfisols has been possible due to the favourable soil temperature and moisture regime. Several authors reported relationships between clay content, silicate clay types with OC content (Pal et al. 2015). Although the 2:1expanding clay minerals provide higher surface area for OC accumulation, the SAT Vertisols, which are dominantly smectitic, even under sub-humid

9.2 Carbon Sequestration in Indian Tropical Soils

Table 9.1 Some selected chemical properties of soils related to OC sequestration

Horizon	OC (%)	pH (1:2)	Clay CEC cmol (p⁺) kg^{-1}	Dominant clay mineral
\multicolumn{5}{l}{Typic Haplusterts as representative of Vertisols of sub-humid moist (SHM) climate[a]}
Ap	1.1[1]	6.3	96	Smectite (partially hydroxy interlayered)
Bw1	0.6	6.3	88	
Bw2	0.6	6.4	90	
Bss1	0.6	6.4	82	
Bss2	0.5	6.5	97	
Bss3	0.1	6.6	102	
\multicolumn{5}{l}{Typic Haplusterts as representative of Vertisols of tropical humid (HT) climate[a]}
Ap	0.9[2]	6.6	51	Mixed mineral (smectite-kaolinite -a 0.7 nm mineral
Bw	0.7	6.4	48	interstratified with hydroxy-interlayered smectite, HIS)
Bss1	0.6	6.8	47	
Bss2	0.5	6.7	46	
Bss3	0.5	6.5	64	
\multicolumn{5}{l}{Vertic Argiudoll as representative of Mollisols of HT climate[b]}
A1	2.0[3]	5.7	36	Mixed mineral (smectite-kaolinite - a 0.7 nm mineral
Bw	1.2	5.7	35	interstratified with hydroxy-interlayered smectite, HIS)
B1t	0.7	5.7	31	
B2t	0.4	6.1	30	
B3t	0.3	6.1	32	
\multicolumn{5}{l}{Typic Haplustalfs as representative of Alfisols of HT climate[b]}
A	1.3[4]	5.7	15	Mixed mineral (smectite-kaolinite - a 0.7 nm mineral
Bw1	1.2	5.3	17	interstratified with hydroxy-interlayered smectite, HIS)
B1t	1.0	5.3	21	
B2t	0.9	5.6	13	
B3t	0.6	5.6	10	
\multicolumn{5}{l}{Ustic Kandihumults as representative of Ultisols of HT climate[c]}
Ap	2.35[5]	4.8	21	Mixed mineral (smectite-kaolinite - a 0.7 nm mineral
Bt1	1.86	4.4	11	interstratified with hydroxy-interlayered vermiculite,
Bt2	1.50	4.5	13	HIV)
Bt3	0.90	4.5	16.	
Bt4	1.11	4.4	14	
Bt5	1.22	4.7	17	
\multicolumn{5}{l}{Pachic Fulvunand as representative of Andosols of HT climate[d]}
A1	12.6[6]	4.5	64	Mixed mineral (hydroxy-interlayered vermiculite) with
A2	10.2	4.5	51	some amount of 0.7 nm mineral interstratified with
A3	6.7	4.6	45	hydroxy-interlayered vermiculite, HIV)
Bw	3.5	4.8	34	

[a]Adapted from Pal et al.(2009)
[b]Adapted from Bhattacharyya et al. (2005)
[c]Adapted from Chandran et al. (2005)
[d]Adapted from Caner et al. (2000)

(continued)

Table 9.1 (continued)
[1]OC of 0.67% as weighted mean (WM) value in the 0–100 cm of the profile
[2]OC of 0.65% as WM value in the 0–100 cm of the profile
[3]OC of 0.94% WM value in the 0–100 cm of the profile
[4]OC of 1.03% as WM value in the 0–100 cm of the profile
[5]OC of 1.45% as WM value in the 0–100 cm of the profile
[6]OC of 8.29% as WM value in the 0–100 cm of the profile

moist bio-climate supporting two agricultural crops in a year, are not OC enriched (~ 1%) (Table 9.1) (Pal et al. 2003). On the other hand, zeolitic but acidic Vertisols (Typic Haplusterts, Pal et al. 2009) and Mollisols (Vertic Haplustoll/Argiudoll) of HT climate show a similar value or an increase in OC content ~1% and 2.0%, respectively (Bhattacharyya et al. 2006), whereas the acidic but zeolitic Alfisols of HT climate also show a higher OC content (> 1%) (Table 9.1) (Bhattacharyya et al. 2005; Lal 2000). It is worth mentioning here that HT Vertisols, Mollisols and Alfisols are acidic and their CEC values are <50 cmol (p^+) kg^{-1} (Table 9.1) because of the dominant presence of kaolin (a 0.7 nm mineral interstratified with hydroxy-interlayered smectite, HIS). The dominant presence of kaolin in these HT soils clearly suggests that smectite has been profusely transformed to HIS and then to kaolin (Bhattacharyya et al. 1993), which ultimately has also become a favourable substrate for enhanced sequestration of OC like discrete smectite, the dominant clay minerals of SAT Vertisols (Pal et al. 2009). Formation of acidic Vertisols, Alfisols and Mollisols in presence of soil modifiers like Ca-zeolites suggests that Ca-zeolite is an important factor in OC sequestration as it has high CEC (100–300 cmol (p^+) kg^{-1}) and a large surface area. The Ca-zeolites (as soil modifier) prevent complete transformation of smectite to kaolinite by maintaining relatively high base saturation level in acidic pH Vertisols, Mollisols and Alfisols but could not enhance their CEC values beyond 50 cmol (p^+) kg^{-1}.

Non-smectitic and non-zeolitic soils in HT climate of Kerala, Karnataka, Tamil Nadu, Goa, Andaman and Nicobar, and NEH areas (Pal et al. 2014) and also of sub-humid sub-tropical climate in the sub-montane areas of Uttaranchal (Pal et al. 1987a) under forest have also high OC content, sufficient to qualify as Mollisols. In addition, OC addition to Ultisols and Alfisols of HT climate has been possible due to the favourable soil temperature and moisture regime even in presence of dominant amount of kaolin (0.7 nm minerals interstratified with hydroxy-interlayered vermiculite, HIV). The kaolin from vermiculite also shows a high positive balance of OC in non-smectitic and non-zeolitic soils because kaolin (like the other hydroxy-interlayered minerals) often shows a relatively high value of CEC ~30 cmol (p^+) kg^{-1} (Ray et al. 2001) than that of well crystalline kaolinite. Therefore, besides the dominating effect of humid and sub-humid climate in cooler winter months with profuse vegetation, the soil substrate quality is, however, more influenced by hydroxy-interlayered clay minerals than by discrete smectite, vermiculite and kaolinite to effect OC sequestration.

Clay minerals interstratified with hydroxy-interlayered 2:1 clay minerals, are very common in Vertisols, Alfisols, Mollisols and Ultisols of HT climate (Pal et al.

2014) and also in Mollisols of sub-humid sub-tropical climate (Pal et al. 1987a). Many such acid soils of HT climate have low amount of KCl extractable acidity (due to H$^+$ and Al^{3+} ions) and kaolinitic mineralogy class is assigned to them based on their clay CEC and ECEC values which are less than 16 and 12 cmol (p$^+$) kg^{-1}, respectively (Smith 1986). But it is interesting to note that their total acidity determined by BaCl$_2$-TEA (Peech et al. 1947; Jackson 1973) shows a much higher values than determined by 1 N KCl. The CEC of soils and their clay CEC estimated based on the total acidity plus the sum of bases by 1 N NH$_4$OAc (pH 7) indicates a value much greater than 16 and 12, respectively (See Chap. 7, Table 7.1) (Chandran et al. 2005). The greater values for both soil CEC and clay CEC suggests their mixed mineralogy class that provides an adequate surface area for enhanced C sequestration. Determination of CEC and ECEC using total acidity by BaCl$_2$-TEA plus bases by 1 N NH$_4$OAc (pH 7) is a straightforward and simple analytical method to infer the real mineralogy class. The unique role of hydroxy-interlayered clay minerals in OC sequestration in soils is amply cleared from the OC (WM in the 0–100 cm of the soil profiles) and clay CEC values (based on soil CEC, NH$_4$OAC, pH 7) (Table 9.1), which show an increase in OC but a decrease in clay CEC value due to higher rate of hydroxy-interlayering in the interlayers of 2:1 layer silicates from Vertisols to Ultisols. The OC values are 0.67% for SHM Vertisols, 0.65% for HT Vertisols, 0.94% for HT Mollisols, 1.03% for HT Alfisols and 1.45% for HT Ultisols whereas the corresponding clay CEC values are >50 for SHM Vertisols, < 50 for HT Vertisols, <40 for Mollisols and < 25 cmol (+) kg^{-1} for Alfisols and Ultisols (Table 9.1).

It is interesting to note the presence of Andisols under the HT climate of the Nilgiri Hills, southern India (Caner et al. 2000). These soils are clayey, highly acidic, and enriched with OC ranging from 8.0–14.0% in the 0–30 cm of the profile and its WM value in the 0-100 cm profile is 8.29% (Table 9.1). These Andisols have dominant hydroxy-interlayered vermiculite (HIV) clay minerals as evident from their moderate to high clay CEC > 30 but <100 cmol (+) kg^{-1} (Caner et al. 2000) (Table 9.1). However, these authors explained their OC enrichment is due to metal-humus complexes as explained by Oades (1989) and Chatterjee et al. (2013). Under acidic soil condition, Fe and Al oxides become appropriate source of positively charged metal cations for metal-humus complexes (Caner et al. 2000). However, under similar climate with Fe-Al oxide rich parent materials of the Nilgiri Hills of southern India, occurrence of acidic Mollisols is very common (Krishnan et al. 1996; Natarajan et al. 1997). OC enrichment in soils is also ascribed to the presence of poorly crystalline layer silicates (Datta et al. 2015) and Fe and Al oxides (Chatterjee et al. 2013). Based on the hypothesis of metal-humus complex, Caner et al. (2000) proposed the formation of non-allophanic Andosols on non-volcanic materials consisting of Fe and aluminium oxides even in abundant presence of clay HIV (Table 9.1). The mechanism of the enrichment of OC in HT soils by the hydroxy-interlayered materials (HIM) in vermiculite and smectite is still not fully highlighted though Fe,Al and Mg hydroxides released during the acidic tropical weathering of soils also remain as positive cations ([Fe$_3$ (OH)$_6$]$^{3+}$,[Al$_6$(OH)$_{15}$]$^{3+}$, [Mg$_2$Al(OH)$_6$]$^+$, [Al$_3$(OH)$_4$]$^{5+}$, etc. Barnhisel and Bertsch 1989) before they enter the interlayer spaces of 2:1 layer silicate minerals at pH much below 8.3 (Jackson

1964). Moderately acidic conditions are optimal for hydroxy-Al interlayering of vermiculite and smectite and the optimum pH for interlayering in smectite and vermiculite is 5.0–6.0 and 4.5–5.0 respectively (Rich 1968) as small hydroxyl ions are most likely to be produced at low pH (Rich 1960). Ray et al. (2006) and Thakare et al. (2013) used 0.25 N EDTA solutions (pH 7.0) to remove the HIM materials from the fine-clay smectites of Indian Vertisols to determine the layer charge of the cleaned clays and observed the removal of HIM materials by the EDTA solutions was almost complete. Although the methods to determine poorly crystalline layer silicates and Fe and Al oxides and HIM are available, a special research attention is warranted on the nature and properties of hydroxy-interlayered materials to establish their positive role (if any) in enriching OC in HT soils.

9.3 Acidity, Al Toxicity, Lime Requirement and Phosphate Fixation in Soils of Indian Tropical Environments

It is well documented that soil acidity is a major constraint to crop production because the exchangeable acidity of HT soils, which is mainly controlled by the significant contribution of Al_3^+ ions. To ameliorate the soil acidity, liming is generally recommended for strong to moderately acid soils of tropical India. However, crop response to liming remains anomalous. In some moderately acidic Alfisols of the Western Ghats, crop response to liming is not observed (Kadrekar 1979) due to the presence of Ca-zeolites in these soils (Pal et al. 2012a). Although the aluminium toxicity to plants is a constraint associated with overall low nutrient reserves in Oxisols, Ultisols and Dystropepts (Sanchez and Logan 1992), reports on Al toxicity to crops in Indian acidic soils are scarce. Ultisols and Dystropepts of Kerala, Goa, Tamil Nadu, Karnataka and NEH are strongly to moderately acidic, and liming is often recommended to correct soil acidity and improve nutrient availability (ICAR-NAAS 2010). Liming is recommended to bring exchangeable aluminium level in the soil to <1 mg kg^{-1} (Sanchez 1976) for maintaining sustained productivity because it improves the general nutrient availability. The KCl extractable aluminium is used for liming tropical acidic pH soils (Oates and Kamprath 1983). It is, however, intriguing that despite equally strong acidic Ultisols and Dystropepts of NEH, Ultisols of Kerala, and Alfisols of Goa indicate varying KCl exchange acidity. It is less than 1 cmol (p+) kg^{-1} for the Ultisols of Kerala and Meghalaya, and Alfisols of Goa whereas in the Ultisols of Arunachal Pradesh, Assam, Mizoram, Nagaland, Tripura and Dystrochrepts of Manipur, it ranges from 3 to 6 cmol (p+) kg^{-1} (Pal et al. 2014). Accordingly, the lime requirement (LR) of these acidic soils varies widely; lime requirement is around 1 t lime ha^{-1} for Ultisols of Meghalaya, and for Dystrochrepts and Ultisols of other states of NEH, it ranges from 4 to 12 t ha^{-1} (Pal et al. 2014). A low LR (< 1 t ha^{-1}) may be expected for zeolitic soils that contain higher amount of soluble Ca-ions from dissolution of Ca-zeolite but for the gibbsite containing Ultisols (Kerala, Meghalaya) and Alfisols (Goa), the reason for low KCl

9.3 Acidity, Al Toxicity, Lime Requirement and Phosphate Fixation in Soils of Indian...

exchangeable acidity and LR, it is still not clear. It is interesting to note that highly acidic Ultisols of Kerala, Karnataka, Tamil Nadu and NEH and moderately acidic Alfisols of Goa, Karnataka and Tamil Nadu have very high $BaCl_2$-TEA extractable acidity in contrast to their low to very low 1 N KCl exchangeable acidity (See Chap. 7, Table 7.1) (Pal et al. 2014). This indicates that Al_3^+ ions released during the humid tropical weathering are trapped as $Al(OH)_2^+$ ion in the interlayers of 2:1 minerals to form hydroxy interlayered minerals (Bhattacharyya et al. 2000b). Both smectites and vermiculites act as sinks for aluminium and thus protect the biota from Al toxicity. If Al_3^+ ions were not fixed these cations might have created very high acidity as well as Al-toxicity in soils to render them problematic for agricultural use. Therefore, for soils containing Ca-zeolites and gibbsite, recommendation for LR on the basis of KCl exchangeable acidity needs to be regarded with caution.

Moderate to highly acidic red ferruginous (RF) soils adsorb phosphorus (P) from added phosphate fertilizers. Moreover, large doses of P are required to get desired crop response even when crop requirement of P is relatively low (Datta 2013). Clays rich in 1:1 lattice mineral may contribute to P adsorption in highly weathered soils of humid tropical climate, especially at low soil pH, when the activity of iron and aluminium is high. Thus, the Fe and Al-oxy-hydroxides ordinarily present as fine coatings on surfaces of silicate clay minerals (Haynes, 1983) in RF soils, have phosphate fixing ability and can adsorb large amounts of added P. However, these minerals can adsorb negatively charged phosphate ions only when they remain as cations in highly acidic medium. Many Alfisols and Ultisols of India are highly acidic and their KCl pH values remain close to equal or greater than water pH (See Chap. 7, Table 7.1) (Bhattacharyya et al. 2000b; Chandran et al. 2004, 2005), indicating the presence of gibbsite and/or poorly crystalline materials (Smith 1986). This indicates that gibbsite and/or sesquioxides showing a positive ΔpH could be a better substrate to absorb negatively charged phosphate ions. It is evident from the continuous increment in yield of rice up to 120 kg $P_2O_5 ha^{-1}$ in gibbsitic soils of Meghalaya (Datta 2013). Therefore, the highest surface area of 2:1 expanding silicate clay minerals and/or Fe and Al-oxy-hydroxides with no positive sites, have little role in the adsorption of added negatively charged phosphate ions in mild to moderately acidic soils. Therefore, determination pH in both water and 1 N KCl solution is essential and the KCl value greater than water will be a simple analytical method to evaluate the soils for their phosphate fertilization.

Mildly acidic SAT Alfisols contain silt and clay size hydroxy interlayered vermiculite (HIV) and hydroxy-interlayered dioctahedral smectite (HIS) (Pal and Deshpande 1987). Fine clay smectites of SAT Vertisols are partially hydroxy-interlayered (Pal et al. 2012a). Such hydroxy interlayering is not the contemporary pedogenic process, because in the prevailing mild acidic pH conditions in Alfisols and slight to moderate alkaline pH conditions of SAT Vertisols, the Fe and Al-oxy-hydroxides do not exist as positively charged cations (Pal et al. 2012b). The observed P adsorption is thus related to the formation of Ca–P in SAT Alfisols and Vertisols that contain ~1% (Pal et al. 2014) and > 5% $CaCO_3$ (Pal et al. 2012a), respectively as calcification is one of the contemporary pedogenetic processes in SAT soils (Pal et al. 2000b). Thus, the simple carbonate clay data would be of much help in resolving the problem of phosphate fixation in SAT soils in general.

9.4 K Release from Biotite Mica and Adsorption of Nitrogen and Potassium by Vermiculite of Indian Tropical Soils

Although the major soil types of tropical India (IGP, cracking clay and red ferruginous soils), endowed with fine-grained micas, known as natural K suppliers to plants, crop response to K fertilizers has been anomalous. Although the silt and clay size muscovite and biotite co-exist in soils, the rate of K release and crop response to K are related primarily to the presence of biotite only while muscovite remains as an inert source K in soils justifying the crop response to K fertilizers (Pal et al. 2001). Due to profuse weathering of biotite micas in HT climate, weathered product like di-octahedral vermiculite is formed (Pal et al. 1987b). In contrast, in SAT environments biotite micas alter to trioctahedral vermiculite and smectite (Pal et al. 1987b, 1989). The rate of K release is a function of the content of biotite mica in soils and thus the K release is highest in biotite enriched IGP soils followed by ferruginous and cracking clay soils (Pal and Durge 1993). The lowest rate of K release is observed in muscovite containing Ultisols of HT climate. Due to low content of biotite micas, soils of HT climate have very low extractable K (< 100 ppm) (Bhattacharyya et al. 1994), and SAT ferruginous and IGP soils have high to very high extractable K (100–350 ppm) as they are enriched with biotite micas (Pal et al. 2001). SAT cracking clay soils also show a very high extractable K (> 300 ppm) as they contain both biotite and zeolite (Pal et al. 2006, 2013). Although many a times, the importance of extractable K does not get proper attention for K management of soils, the amount of extractable K is definitely the most simple and scientific way to comprehend the abundance of available K due to the presence of biotite mica in major soil types of India.

Due to weathering of biotite micas, all the three soil types have discrete vermiculite and low charge vermiculite or high charge smectite, which are responsible for the adsorption of both NH_4 and K ions. It is a well-established fact that only vermiculite and/or hydroxy-interlayered vermiculite (HIV) and low charge vermiculite are capable of K adsorption (Pal et al., 2012a, 2014) Therefore, it would be prudent to attribute the observed NH_4-N and K adsorption to the presence of only vermiculite and low charge vermiculite (Pal et al. 2012a, 2014) in tropical soils of India (Pal and Durge 1987, 1989; Pal et al. 1993; Sahrawat 1995) and elsewhere (Fox 1982). Di-octahedral smectites in SAT ferruginous and Vertisols have no selectivity for non-hydrated monovalent cations such as K^+, because of their low layer charge (Brindley 1966). NH_4^+ ion, also a non-hydrated monovalent cation with almost the same ionic radius as K, is not expected to be fixed in the interlayers of low charge di-octahedral smectites. It is equally difficult to understand the NH_4 and K ion fixing capacity of illites, because they do not expand on being saturated with divalent cations (Sarma 1976). Presence of the K and NH$_4$ fixing minerals (vermiculite and/ or hydroxy-interlayered vermiculite and low charge vermiculite) in soils can be easily identified from the proportion of extractable K occupying in the CEC of soils [extractable K in cmol (p$^+$) kg^{-1}/ CEC in cmol (p$^+$) kg^{-1} x 100]. It is observed that extractable K has a significant amount in CEC for soils of HT climate (> 1 < 6%) and also in IGP and ferruginous soils (> 2 < 6%) (Pal et al. 2010; Murthy et al. 1982), and it has the least amount (>1 < 2%) in smectite dominated cracking clay soils (Pal et al. 2003). Careful observation of such simple data base helps to understand the extent of K fixation/adsorption of added K for better K management of soils.

References

Barnhisel RI, Bertsch PM (1989) Chlorites and hydroxy-interlayered vermiculites and smectite. In: Dixon JB, Weed SB (eds) Minerals in soil environments, Second Edition. Soil Science Society of America Book Series (Number 1). Wisconsin, USA, pp 729–788

Batjes H (1996) Total carbon and nitrogen in the soils of the world. European J Soil Sci 47:151–163

Bhattacharyya T, Pal DK, Deshpande SB (1993) Genesis and transformation of minerals in the formation of red (Alfisols) and black (Inceptisols and Vertisols) soils on Deccan basalt in the western Ghats, India. J Soil Sci 44:159–171

Bhattacharyya T, Sen TK, Singh RS, Nayak DC, Sehgal JL (1994) Morphology and classification of Ultisols with Kandic horizons in northeastern region. J Indian Soc Soil Sci 42:301–306

Bhattacharyya T, Pal DK, Velayutham M, Chandran P, Mandal C (2000a) Total carbon stock in Indian soils: issues, priorities and management. In: Land resource management for food and environmental security. Soil Conservation Society of India, New Delhi, pp 1–46

Bhattacharyya T, Pal DK, Srivastava P (2000b) Formation of gibbsite in presence of 2:1 minerals: an example from Ultisols of Northeast India. Clay Miner 35:827–840

Bhattacharyya T, Pal DK, Chandran P, Ray SK (2005) Land-use, clay mineral type and organic carbon content in two Mollisols–Alfisols–Vertisols catenary sequences of tropical India. Clay Res 24:105–122

Bhattacharyya T, Pal DK, Lal S, Chandran P, Ray SK (2006) Formation and persistence of Mollisols on Zeolitic Deccan basalt of humid tropical India. Geoderma 136:609–620

Bhattacharyya T, Pal DK, Chandran P, Ray SK, Mandal C, Telpande B (2008) Soil carbon storage capacity as a tool to prioritise areas for carbon sequestration. Curr Sci 95:482–494

Bhattacharyya T, Sarkar D, Sehgal JL, Velayutham M, Gajbhiye KS, Nagar AP, Nimkhedkar SS (2009) Soil taxonomic database of India and the states (1:250,000 scale), NBSSLUP Publ. 143, NBSS&LUP, Nagpur, India, p 266

Brindley GW (1966) Ethylene glycol and glycerol complexes of smectites and vermiculites. Clay Miner 6:237–259

Caner L, Bourgeon G, Toutain F, Herbillon AJ (2000) Characteristics of non-allophanic Andisols derived from low-activity clay regoliths in the Nilgiri Hills (southern India). European J Soil Sci 51:553–563

Chandran P, Ray SK, Bhattacharyya T, Dubey PN, Pal DK, Krishnan P (2004) Chemical and mineralogical characteristics of ferruginous soils of Goa. Clay Res 23:51–64

Chandran P, Ray SK, Bhattacharyya T, Srivastava P, Krishnan P, Pal DK (2005) Lateritic soils of Kerala, India: their mineralogy, genesis and taxonomy. Aust J Soil Res 43:839–852

Chatterjee D, Datta SC, Manjaiah KM (2013) Clay carbon pools and its relationship with short range order minerals: avenues for climate change? Curr Sci 105:1404–1410

Datta M (2013) Soils of north-eastern region and their management for rain-fed crops. In: Bhattacharyya T, Pal DK, Sarkar D, Wani SP (eds) Climate change and agriculture. Studium Press, New Delhi, pp 19–50

Datta SC, Takkar PN, Verma UK (2015) Assessing stability of humus in soils from continuous rice-wheat and maize-wheat cropping systems using kinetics of humus desorption (2015). Commun Soil Sci Plant Anal 46:2888–2900. https://doi.org/10.1080/00103624.2015.1104334

Fox RL (1982) Some highly weathered soils of Puerto Rico, 2.Chemical properties. Geoderma 27:139–176. https://doi.org/10.1016/0016-7061(82)90049-0

Haynes RJ (1983) Effect of lime and phosphate applications on the adsorption of phosphate, sulphate, and molybdate by a Spodosol. Soil Sci 135:221–226

ICAR-NAAS (Indian Council of Agricultural Research-National Academy of Agricultural Sciences) (2010) Degraded and wastelands of India-status and spatial distribution. Indian Council of Agricultural Research and National Academy of Agricultural Sciences. Published by the Indian Council of Agricultural Research, New Delhi, p 56

Jackson ML (1964) Chemical composition of soils. In: Bear FE (ed) Chemistry of the soil. Oxford and IBH Publishing Co., Calcutta, pp 71–141

Jackson ML (1973) Soil chemical analysis. Prentice Hall of India Pvt Ltd, New Delhi

Kadrekar SB (1979) Utility of basic slag and liming material in lateritic soils of Konkan. Indian J Agron 25:102–104

Krishnan P, Venugopal KR, Sehgal J (1996) Soil Resources of Kerala for land use planning. NBSS Publ. 48b (Soils of India Series 10), National Bureau of Soil Survey and Land Use Planning, Nagpur, India, 54 pp + 2 sheets of soil map on 1:500,000 scale

Lal S (2000) Characteristics, genesis and use potential of soils of the Western Ghats, Maharashtra. Ph. D thesis, Dr. PDKV, Akola, Maharashtra, India

Murthy RS, Hirekerur LR, Deshpande SB, Venkat Rao BV (eds) (1982) Benchmark soils of India. National Bureau of Soil Survey and Land Use Planning, Nagpur, p 374

Natarajan A, Reddy PSA, Sehgal J, Velayutham M (1997) Soil Resources of Tamil Nadu for land use planning. NBSS Publ. 46b (Soils of India Series), National Bureau of Soil Survey and Land Use Planning, Nagpur, India, 88 pp + 4 sheets of soil map on 1:500,000 scale

Oades JM (1989) An introduction to organic matter in mineral soils. In: Dixon JB, Weed SB (eds) Minerals in soil environments, 2nd edn. Soil Science Society of America, Madison, pp 89–159

Oates KM, Kamprath EJ (1983) Soil acidity and liming: effects of the extracting solution cations and pH on the removal of aluminium from acid soils. Soil Sci Soc Am J 47:686–690

Pal DK, Deshpande SB (1987) Genesis of clay minerals in a red and black complex soils of southern India. Clay Res 6:6–13

Pal DK, Durge SL (1987) Potassium release and fixation reactions in some benchmark Vertisols of India in relation to their mineralogy. Pedologie (Ghent) 37:103–116

Pal DK, Durge SL (1989) Release and adsorption of potassium in some benchmark alluvial soils of India in relation to their mineralogy. Pedologie (Ghent) 39:235–248

Pal DK, Durge SL (1993) Potassium release from clay micas. J Indian Soc Soil Sci 41:67–69

Pal DK, Deshpande SB, Sehgal JL (1987a) Development of soils in quaternary deposits of North India. Indian J Earth Sci 14:329–334

Pal DK, Deshpande SB, Durge SL (1987b) Weathering of biotite in some alluvial soils of different agro climatic zones. Clay Res 6:69–75

Pal DK, Deshpande SB, Venugopal KR, Kalbande AR (1989) Formation of di- and trioctahedral smectite as an evidence for paleoclimatic changes in southern and central peninsular India. Geoderma 45:175–184

Pal DK, Deshpande SB, Durge SL (1993) Potassium release and adsorption reactions in two ferruginous soils (polygenetic) soils of southern India in relation to their mineralogy. Pedologie (Ghent) 43:403–415

Pal DK, Bhattacharyya T, Deshpande SB, Sarma VAK, Velayutham M (2000a) Significance of minerals in soil environment of India, NBSS Review Series 1. NBSS&LUP, Nagpur, p 68

Pal DK, Dasog GS, Vadivelu S, Ahuja RL, Bhattacharyya T (2000b) Secondary calcium carbonate in soils of arid and semi-arid regions of India. In: Lal R, Kimble JM, Eswaran H, Stewart BA (eds) Global climate change and pedogenic carbonates. Lewis Publishers, Boca Raton, pp 149–185

Pal DK, Srivastava P, Durge SL, Bhattacharyya T (2001) Role of weathering of fine grained micas in potassium management of Indian soils. Applied Clay Sci 20:39–52

Pal DK, Bhattacharyya T, Ray SK, Bhuse SR (2003) Developing a model on the formation and resilience of naturally degraded black soils of the peninsular India as a decision support system for better land use planning. NRDMS, Department of Science and Technology (Govt. of India) Project Report, NBSSLUP (ICAR), Nagpur, p 144

Pal DK, Nimkar AM, Ray SK, Bhattacharyya T, Chandran P (2006) Characterisation and quantification of micas and smectites in potassium management of shrink–swell soils in Deccan basalt area. In: Benbi DK, Brar MS, Bansal SK (eds) Balanced fertilization for sustaining crop productivity. Proceedings of the International Symposium held at PAU, Ludhiana, India, 22–25 Nov' 2006 IPI, Switzerland, pp 81–93

Pal DK, Bhattacharyya T, Chandran P, Ray SK, Satyavathi PLA, Durge SL, Raja P, Maurya UK (2009) Vertisols (cracking clay soils) in a climosequence of peninsular India: evidence for Holocene climate changes. Quatern Int 209:6–21

Pal DK, Lal Sohan, Bhattacharyya T, Chandran P, Ray SK, Satyavathi PLA, Raja P, Maurya, UK, Durge SL, Kamble GK (2010) Pedogenic thresholds in benchmark soils under rice-wheat cropping system in a climosequence of the Indo-Gangetic alluvial plains. Final Project Report, Division of Soil Resource Studies, NBSS&LUP (ICAR), Nagpur, p 193

Pal DK, Wani SP, Sahrawat KL (2012a) Vertisols of tropical Indian environments: pedology and edaphology. Geoderma 189-190:28–49

Pal DK, Bhattacharyya T, Sinha R, Srivastava P, Dasgupta AS, Chandran P, Ray SK, Nimje A (2012b) Clay minerals record from late quaternary drill cores of the Ganga Plains and their implications for provenance and climate change in the Himalayan foreland. Palaeogeogr Palaeoclimatol Palaeoecol 356–357:27–37

Pal DK, Wani SP, Sahrawat KL (2013) Zeolitic soils of the Deccan basalt areas in India: their pedology and edaphology. Curr Sci 105:309–318

Pal DK, Wani SP, Sahrawat KL, Srivastava P (2014) Red ferruginous soils of tropical Indian environments: a review of the pedogenic processes and its implications for edaphology. Catena 121:260–278. https://doi.org/10.1016/j.catena2014.05.023

Pal DK, Wani SP, Sahrawat KL (2015) Carbon sequestration in Indian soils: present status and the potential. Proc Natl Acad Sci, Biol Sci (NASB), India 85:337–358. https://doi.org/10.1007/s40011-014-0351-6

Peech M, Alexander LT, Dean LA, Reed JF (1947) Methods of soil analysis and soil fertility investigations. U S Department of Agriculture, Circular No.752

Ray SK, Chandran P, Durge S.L (2001) Soil taxonomic rationale: kaolinitic and mixed mineralogy classes of highly weathered ferruginous soils. Abstract, 66th Annual Convention and National Seminar on "Developments in soil science" of the Indian Society of Soil Science, Udaipur, Rajasthan, pp 243–244

Ray SK, Chandran P, Bhattacharyya T, Pal DK (2006) Determination of layer charge of soil smectites by alkylammonium method: effect of removal of hydroxy-interlayering. Abstract, 15th annual convention and National Symposium on"clay research in relation to agriculture, environment and industry"of the clay minerals Society of India. BCKVV, Mohanpur, p 3

Rich CI (1960) Aluminum in interlayers of vermiculite. Soil Sci Soc Amer Proc 24:26–32

Rich CI (1968) Hydroxy-interlayering in expansible layer silicates. Clay Clay Miner 16:15–30

Sahrawat KL (1995) Fixed ammonium and carbon–nitrogen ratios of some semi-arid tropical Indian soils. Geoderma 68:219–224

Sanchez PA (1976) Properties and Management of Soils in the tropics. John Wiley and Sons, New York

Sanchez PA, Logan TJ (1992) Myths and science about the chemistry and fertility of soils in the tropics. In: Lal R, Sanchez PA (eds) Myths and science of soils of the tropics. SSSA Special Publication Number 29. SSSA, Inc and ACA, Inc, Madison, pp 35–46

Sarma VAK (1976) Mineralogy of soil potassium. Bull Indian Soc Soil Sci 10:66–77

Smith GD (1986) The Guy Smith interviews: rationale for concept in soil taxonomy. SMSS Technical Monograph, 11. SMSS, SCS, USDA, USA

Srivastava P, Pal DK, Aruche KM, Wani SP, Sahrawat KL (2015) Soils of the indo-Gangetic Plains: a pedogenic response to landscape stability, climatic variability and anthropogenic activity during the Holocene. Earth Sci Rev 140:54–71. https://doi.org/10.1016/j.earscirev.2014.10.010

Thakare PV, Ray SK, Chandran P, Bhattacharyya T, Pal DK (2013) Does sodicity in Vertisols affect the layer charge of smectites? Clay Research 32:76–90

Velayutham M, Pal DK, Bhattacharyya T (2000) Organic carbon stock in soils of India. In: Lal R, Kimble JM, Stewart BA (eds) Global climate change and tropical ecosystems. Lewis Publishers, Boca Raton, Fl, pp 71–96

Chapter 10
A Critique on Degradation of HT and SAT Soils in View of Their Pedology and Mineralogy

Abstract In the present critique, discussion shall remain restricted to the physical (soil loss due to water erosion) and natural chemical degradation caused by two extreme climatic conditions i.e. semi-arid (SAT) and humid tropical (HT) climates. Past research reports indicated that the red ferruginous (RF) of both HT and SAT climates are physically and chemically degraded to a considerable extent. However, newly acquired research data on the rate of top soil formation and non-kaolinitic mineralogy class indicate that the extent of physical and chemical degradation of RF soils of HT climate (especially Ultisols, and acidic Alfisols, Mollisols and Inceptisols with clay enriched B horizons) and also of SAT climate (mainly Alfisols) is not at all in an alarming stage. Proper understanding of the operative pedogenic processes and the determination of the exact mineralogy class are of mandatory requirements for suggesting/deciding the extent of degradation of HT and SAT soils. Finally, this critique provides a deductive check on the inductive reasoning on the physical and chemical degradation of RF soils of both HT and SAT climates and also highlights the importance of the basic pedogenic processes and determination of the mineralogy class by simple analytical methods while deciding the extent of degradation of SAT and HT soils.

Keywords Soil degradation · Pedogenesis · Mineralogy

ICAR-NAAS (2010) considered water erosion to be the most widespread form of soil degradation in the Indian sub-continent, affecting an area of 73.27 m ha area in all agro-climatic zones. Soils under HT climate of NEH (north-eastern hills) and southern peninsular areas are the most affected by water erosion based on assumption that soil erosion <10 t ha^{-1} yr^{-1} (using the empirical Universal Soil Loss Equation) does not significantly affect soil productivity and thus they are not degraded (ICAR-NAAS 2010). Such assumption on soil loss may not be prudent in soils under HT climate, because of the dominance of matured soils such as Ultisols, Alfisols, Mollisols, and Inceptisols with clay enriched B horizons in NEH areas, Kerala, Goa, Maharashtra, Madhya Pradesh, Karnataka and Andaman and Nicobar Islands (Pal et al. 2014). These soils do exhibit a higher rate of soil formation due to the active pedogenic processes continuing for several years (Fig. 10.1a-c).

Fig. 10.1 Representative clayey Ultisols (**a**), Alfisols (**b**) and Mollisols (**c**) of HT and clayey Alfisols of SAT(**d**) climates (Photos, courtesy-author)

The role of the major pedogenic processes are evident through the rich addition of C by litter falls and its accumulation as soil organic matter under adequate vegetation and climate, translocation of clay particles (to form clay enriched B horizons) and transformation of 1.4 nm minerals to 0.7 nm minerals (kaolin). In dry and cold environments, the rate of soil formation varies from <0.25 mm yr^{-1} to >1.5 mm yr^{-1} in humid and warm environments (Kassam et al. 1992). So even if the rate of 2 mm yr^{-1} is taken for top soil formation in HT climate, then it would amount to 29 t ha^{-1} (a value much higher than the assumed value of 10 t ha^{-1} as the threshold value of soil degradation by water erosion) when the weight of hectare-furrow slice (15 cm depth) is taken as 2.2×10^6 kg. This rate of soil formation justifies the existence of matured soils of the NEH areas (Bhattacharyya et al. 2007a) and also proves that the rate of soil loss by water erosion from Ultisols, Alfisols, Mollisols and Inceptisols is minimal (Pal et al. 2014). This fact is quite compatible with the widespread occurrence of Ultisols, Alfisols, Mollisols and clay enriched Inceptisols on a stable landscape under adequate vegetation and profuse weathering of parent material in HT climate. However, in RF soils (mainly Entisols) on higher slopes (ridges, scarps and terraces) under low vegetation with only shrubs and bushes, soil development is greatly hampered by the severe soil loss due to water erosion. Soil loss is also evident in other soils that have less vegetative cover.

Considering the soil loss >10 t ha^{-1} yr^{-1} as the threshold for soil degradation, SAT soils of Indian states are shown to suffer soil loss due to erosion (ICAR-NAAS 2010). The RF soils of SAT dominantly belong to Alfisols (Fig. 10.1d), and the other soil orders are Inceptisols, Entisols and Mollisols (Pal et al. 2014). If the rate of soil formation in dry environments is taken at <0.25 mm yr^{-1} (Kassam et al. 1992), SAT Alfisols would gain soil at least 3.67 t ha^{-1} yr^{-1}, which is much lower than the assumed value of 10 t ha^{-1} yr^{-1} (ICAR-NAAS 2010). However, both short-term and long-term hydrological studies on small agricultural watershed on Alfisols at the ICRISAT Center, Patancheru, India indicate an average soil loss from SAT Alfisols under traditional system is around 3.84–4.62 t ha^{-1} yr^{-1} (Pathak et al. 1987, 2013). The higher soil loss on the SAT Alfisols under traditional management is due to clay enriched B horizons with substantial amount of smectite silicate clay mineral in the subsoils (Chandran et al. 2009; Pal et al. 1989), which is reflected in COLE >0.06 (Bhattacharyya et al. 2007b), and reduced sHC (Pathak et al. 2013) due to formation of pedogenic CaCO$_3$ (PC) and concomitant development of subsoil sodicity (ESP) (Chandran et al. 2011; Pal et al. 2000, 2013). Such physical and mineralogical environments in the subsoil restrict vertical movement of water in the soil profile, resulting in greater soil loss from the SAT Alfisols through overland lateral flow of water. To reduce unwarranted soil loss and also to improve the sustainability of the SAT Alfisols, improved system of management developed by the ICRISAT (Pathak et al. 1987) minimizes soil loss to nearly 1 t ha^{-1}yr^{-1} and simultaneously increased crop productivity compared to traditional system. It is hence suggested that due consideration of the rate of soil formation during the active pedogenic processes that lead to formation of matured soil profiles, is mandatory while assessing the status of physical degradation like water erosion in HT and SAT red ferruginous soils. Moreover, the threshold value of soil loss by water erosion as sign of degradation, should strictly be based on experimental results rather than assuming an arbitrary value of >10 t ha^{-1} yr.$^{-1}$ (ICAR-NAAS 2010).

Acid soils develop under HT climatic environment with a loss of some major nutrient elements, and thus the development of acidity in soils is considered to be a sign of chemical degradation. While estimating the area affected by acidity, only soils with strong (pH < 4.5) and moderate acidity (pH 4.5–5.5) were considered (ICAR-NAAS 2010). Soils of HT climate in the states of Kerala, Goa, Karnataka, Tamil Nadu and NEH areas are strong to moderately acidic Alfisols, Ultisols and Mollisols (Pal et al. 2014); and their further weathering in HT climate would not finally end up in pure kaolinite (0.7 nm mineral) or gibbsite dominated Ultisols but with kaolin (0.7 nm mineral interstratified with hydroxy-interlayered 1.4 nm minerals) with considerable amount of layer silicate minerals (Pal et al. 2012, 2014). This is also reflected in silica: alumina and silica: sesquioxides ratios of the Ultisols of Kerala. The ratios $SiO_2:R_2O_3$ (1.4–5.0) and $SiO_2:Al_2O_3$ (1.8–6.0) indicate the siliceous nature of these soils (Chandran et al. 2005; Varghese and Byju 1993), suggesting an incomplete desilication process. The amount of SiO_2 and its molar ratios are comparable with some of the Oxisols reported from Puerto Rico (Jones et al. 1982), Brazil (Buurman et al. 1996; Muggler 1998), and other regions of the World (Mohr et al. 1972). In the acidic Alfisols, Ultisols and Mollisols, the process of desilication no longer operates in present day conditions because the pH of the soils is well below the threshold of ~ 9 (Millot 1970). The OC rich Ultisols have less Al-saturation in surface horizons due to the downward movement of Al as organo-metal complexes or chelates. But in spite of that they have higher base saturation than the sub-surface horizons due to addition of alkaline and alkaline metal cations (Pal et al. 2014) through litter fall (Nayak et al. 1996). The most important among these processes in Ultisols is that there is no desilication and transformation of kaolin to gibbsite. In view of the contemporary pedogenesis, it is difficult to reconcile that Ultisols would ever be weathered to reach the so-called unproductive Oxisols stage with time frame as envisaged by Smeck et al. (1983) and Lin (2011).

In the past many Alfisols and Ultisols of Kerala, Karnataka and Tamil Nadu were assigned the kaolinitic mineralogy class (Bhattacharyya et al. 2009) based on their clay CEC and ECEC values which are less than 16 and 12 cmol (p+) kg^{-1}, respectively (Smith 1986). However, the acidity of many such soils determined by $BaCl_2$-TEA is much higher than that determined by using 1 N KCl (See Chap. 7, Table 7.1). Total acidity (determined by $BaCl_2$-TEA) plus the sum of bases by NH4OAc (pH 7) when taken for calculating the clay CEC in soil control section, indicates a value much greater than >24 (Chandran et al. 2005). Therefore, kaolinitic mineralogy class may be regarded with caution because such OC rich acid soils with dominant kaolin mineral do respond to management interventions advocated by national agricultural research systems (NARS) and support luxuriant forest vegetation, horticultural, cereal crops, tea, coffee and spices (Sehgal 1998). The success of such land uses provides clear evidence that soil kaolin has considerable chemical reactivity, which is capable of creating an effective substrate to support agriculture and other land uses in the Indian tropical soils. Soil kaolin provides the intrinsic ability to regenerate the productivity of these acid soils, thus making them resilient. This suggests that to sustain crop productivity at an enhanced level, large tracts of lands dominated by acid soils can be brought under improved soil, water and nutrient management to help meet the food needs of ever increasing Indian population and elsewhere of the tropical world (Pal et al. 2014: Gilkes and Prakongkep 2016). Therefore, it would not be prudent to class these OC rich acid soils (See Chap. 9, Table 9.1) as chemically degraded soils. At present, the extent of degradation (soil loss and soil acidity) of Ultisols, Alfisols, Mollisols and Inceptisols with clay enriched B horizons is far from any alarming stage. The above discussion clearly highlights the need of understanding the basic pedogenic processes and determination of the mineralogy class by simple but appropriate analytical methods that are generally followed in the field and laboratory while deciding the extent of chemical degradation in acid soils of HT environments.

References

Bhattacharyya T, Ram B, Sarkar D, Mandal C, Dhyani BL, Nagar AP (2007a) Soil loss and crop productivity model in humid tropical India. Curr Sci 93:1397–1403

Bhattacharyya T, Chandran P, Ray SK, Mandal C, Pal DK, Venugopalan MV, Durge SL, Srivastava P, Dubey PN, Kamble GK, Sharma RP, Wani SP, Rego TJ, Pathak P, Ramesh V, Manna MC, Sahrawat KL (2007b) Physical and chemical properties of selected benchmark spots for carbon sequestration studies in semi-arid tropics of India. Global Theme on Agro-ecosystems Report No. 35. Patancheru 502 324, Andhra Pradesh, India: International Crops Research Institute for the Semi-Arid Tropics (ICRISAT), and New Delhi. Indian Council of Agricultural Research (ICAR), India, p 236

Bhattacharyya T, Sarkar D, Sehgal JL, Velayutham M, Gajbhiye KS, Nagar AP, Nimkhedkar SS (2009) Soil taxonomic database of India and the states (1:250, 000 scale), NBSSLUP Publ. 143, NBSS&LUP, Nagpur, India, p 266

Buurman P, Van Lagen B, Velthorst EJ (1996) Manual of soil and water analysis. Backhuys Publishers, Leiden

Chandran P, Ray SK, Bhattacharyya T, Srivastava P, Krishnan P, Pal DK (2005) Lateritic soils of Kerala, India: their mineralogy, genesis and taxonomy. Aust J Soil Res 43:839–852

Chandran P, Ray SK, Durge SL, Raja P, Nimkar AM, Bhattacharyya T, Pal DK (2009) Scope of horticultural land-use system in enhancing carbon sequestration in ferruginous soils of the semi-arid tropics. Curr Sci 97:1039–1046

Chandran P, Ray SK, Sarkar D, Bhattacharyya T, Mandal C, Pal DK, Nagaraju MSS,Ramakrishna YS, Venkateswarlu B, Chary GR, Nimkar AM, Raja P, Maurya U K,Durge SL, Anantwar SG, Patil SV, Gharami S (2011) Soils of Hayatnagar research farm of CRIDA (Hyderabad), Rangareddy District, Andhra Pradesh. NBSS Publ. No.1037, NBSS&LUP, Nagpur, 174 pp + soil map on 1:5000 scale

Gilkes RJ, Prakongkep N (2016) How the unique properties of soil kaolin affect the fertility of tropical soils. Applied Clay Sci 131:100–136

ICAR-NAAS (Indian Council of Agricultural Research- National Academy of Agricultural Sciences) (2010) Degraded and waste lands of India-status and spatial distribution. ICAR-NAAS. Published by the Indian Council of Agricultural Research, New Delhi, p 56

Jones RC, Hundall WH, Sakai WS (1982) Some highly weathered soils of Puerto Rico, 3.Mineralogy. Geoderma 27:75–137

Kassam AH, van Velthuizen GW, Fischer GW, Shah MM (1992) Agro-ecological land resource assessment for agricultural development planning: a case study of Kenya resources data base and land productivity. Land and water development division. Food and Agriculture Organisation of the United Nations and International Institute for Applied System Analysis, Rome

Lin H (2011) Three principles of soil change and pedogenesis in time and space. Soil Sci Soc Am J 75:2049–2070

Millot G (1970) Geology of clays. Springer-Verlag, New York

Mohr ECJ, Van Baren FA, van Schuylenborgh J (1972) Tropical soil-a comprehensive study of their genesis. Mouton, The Hague, The Netherlands

Muggler CC (1998) Polygenetic Oxisols on tertiary surfaces, Minas Gerais, Brazil: soil genesis and landscape development. (PhD thesis) Wageningen Agricultural University, Wageningen

Nayak DC, Sen TK, Chamuah GS, Sehgal JL (1996) Nature of soil acidity in some soils of Manipur. J Indian Soc Soil Sci 44:209–214

Pal DK, Deshpande SB, Venugopal KR, Kalbande AR (1989) Formation of di and trioctahedral smectite as an evidence for paleoclimatic changes in southern and central Peninsular India. Geoderma 45:175–184

Pal DK, Dasog GS, Vadivelu S, Ahuja RL, Bhattacharyya T (2000) Secondary calcium carbonate in soils of arid and semi-arid regions of India. In: Lal et al (eds) Global climate change and pedogenic carbonates. Lewis Publishers, FL, pp 149–185

Pal DK, Wani SP, Sahrawat KL (2012) Vertisols of tropical Indian environments: pedology and edaphology. Geoderma 189-190:28–49

Pal DK, Sarkar D, Bhattacharyya T, Datta SC, Chandran P, Ray SK (2013) Impact of climate change in soils of semi-arid tropics (SAT). In: Bhattacharyya et al (eds) Climate change and agriculture. Studium Press, New Delhi, pp 113–121

Pal DK, Wani SP, Sahrawat KL, Srivastava P (2014) Red ferruginous soils of tropical Indian environments: a review of the pedogenic processes and its implications for edaphology. Catena 121:260–278. https://doi.org/10.1016/j.catena2014.05.023

Pathak P, Singh S, Sudi R (1987) Soil and water management alternatives for increased productivity on SAT Alfisols. Soil conservation and productivity. Proceedings IV International Conference on Soil Conservation, Maracay-Venezuela, November 3–9, 1985, pp 533–550

Pathak P, Sudi R, Wani SP, Sahrawat KL (2013) Hydrological behaviour of Alfisols and Vertisols in the semi-arid zone: implications for soil and water management. Agric Water Manag 118:12–21

Sehgal JL (1998) Red and lateritic soils: an overview. In: Sehgal J, Blum WE, Gajbhiye KS (eds) Red and lateritic soils. Managing red and lateritic soils for sustainable agriculture, vol 1. Oxford and IBH Publishing Co. Pvt. Ltd., New Delhi, pp 3–10

Smeck NE, Runge ECA, Mackintosh EE (1983) Dynamics and genetic modeling of soil system. In: Wilding LP, Smeck NE, Hall GF (eds) Pedogenesis and soil taxonomy: 1. Concepts and interactions. Elsevier, New York, pp 51–81

Smith GD (1986) The Guy Smith interviews: rationale for concept in soil taxonomy, vol 11. SMSS, SCS, USDA, USA. SMSS technical monograph, Washington

Varghese T, Byju G (1993) Laterite soils, Technical Monograph No.1. State Committee on Science, Technology and Environment. Government of Kerala, Kerala

Chapter 11
Anomalous Potassium Release and Adsorption Reactions: Evidence of Polygenesis of Tropical Indian Soils

Abstract Pedological and mineralogical research in recent years indicate that major soil types of tropical Indian environment, experienced climate change from humid to semi-arid in the geological past. Dominant low charge di-octahedral smectite in shrink-swell soils (Vertisols and Vertic intergrades) formed at the expense of plagioclase feldspar in previous humid tropical (HT) climate is being preserved in the present day semi-arid (SA) climate, which favoured the transformation of almost fresh biotite mica to high charge vermiculite. Red ferruginous (RF) soils of southern India though dominated by kaolin formed at the expense of low charge di-octahedral smectite in previous HT climate, preserved kaolin, and favoured almost unweathered biotite mica to transform to high charge smectite or low charge vermiculite in the prevailing SA climate. Presence of kaolin in soils of the Indo-Gangetic Plains (IGP)also indicate its genesis in previous HT climate but these soils in the present SA climate are dominated by clay mica consisting of both muscovite and biotite. In SA climate biotite transforms to high charge smectite or low charge vermiculite. All these soils contain pedogenic calcium, which is formed in the present SA climate. Therefore, major tropical soils have unique combination of non- silicates and layer silicate minerals, which are climate specific. Therefore these soil types are polygenetic and contain di- and trioctahedral mica, high and low charge smectite and kaolin. As biotites are almost fresh to weakly weathered, K release in relation to soil their particle size may or may not follow the pattern of specimen mica. Kaolinites are of no significance in K adsorption /fixation reaction, while vermiculites are converted to mica by layer contraction by K. Low charge di-octahedral smectites do not possess this property and they do not adsorb K selectively unless the charge density is high like in high charge smectite or low charge vermiculite. It is thus envisaged that the polygenetic nature of tropical Indian soils can be comprehended following their K release and adsorption behaviour because of their unique combination layer silicates that control such reactions.

Keywords Polygenetic Indian tropical soils · Potassium release and adsorption reactions · Biotite · High and low charge smectite and vermiculite

11.1 Introduction

With the change in the soil environment, clay mineral assemblages over time gets modified (Jenkins 1985). However, the pedogenic clay minerals of the intermediate weathering stages, when preserved within a paleosols, can be helpful for paleoclimatic interpretation and also in identifying the polygenesis of soils (Pal et al. 1989). Polygenesis in Indian soils is more a rule than an exception (Pal et al. 2000). This is because major soil types of India (red ferruginous soils, IGP soils and Vertisols) are polygenetic in nature in view of their mineral transformations and modifications of soil properties due to climate change from humid to semi-arid. Due to the termination of humid climate these polygenetic soils have preserved clay minerals of both advanced (smectite/kaolinite, Sm/K) and intermediate weathering stages (biotite, trioctahedral vermiculite, low charge trioctahedral vermiculite or high charge trioctahedral smectite, and low charge dioctahedral smectite). The combination of these layer silicate minerals demonstrates a unique K release and adsorption reactions that are described in the following sections. It is hoped that the thorough understanding of K release and adsorption reactions in relation to mineralogy, will be an easy way to follow the implications of polygenesis of soils.

11.2 K Release and Adsorption in Polygenetic IGP Soils

The IGP soils (Alfisols) older than 2500 yr. BP are relict paleosols (Srivastava et al. 1998, 2015, 2016) and have experienced three climatic episodes during the Holocene period. The first one was cold arid followed by warm and humid, and since 3800 yr. BP it is hot semi-arid. In the early stage of soil formation, biotite weathered to biotite-vermiculite mixed-layer minerals and then to trioctahedral vermiculite, and finally to trioctahedral smectite. This type of biotite weathering has been specific to arid climates (Pal et al. 1989). During the later warmer and wetter climate with maximum humidity, around 6500–3800 yr. BP, stability of smectite became ephemeral and it transformed to Sm/K. Due to termination of wetter phase around 3800 yr. BP, Sm/K could be preserved to the present. During the hot semi-arid climate that followed the lapse of the humid climate, transformation of biotite into its weathering products like trioctahedral vermiculite and smectite did continue. These relict soils have been affected by later climatic change to wetter and warmer conditions as evidenced by the formation of Sm/K at the expense of smectite, and thus qualify to be polygenetic soils (Pal et al. 1989; Srivastava et al. 1998). Even for such polygenetic soils, Pal and Durge (1989) reported that the K release from soil mica (following the repeated batch type for Ba-K exchange) increased with a decrease in soils' particle size, and the rate of K release was dependent upon the nature of soil mica, biotite mica in particular (Fig. 11.1) (Pal et al. 2001a). Most of the sand, silt and clay size biotites are not fresh particles but partially altered during the previous wet climates (Fig. 11.2). Therefore, the observed negative correlation between K release

11.2 K Release and Adsorption in Polygenetic IGP Soils

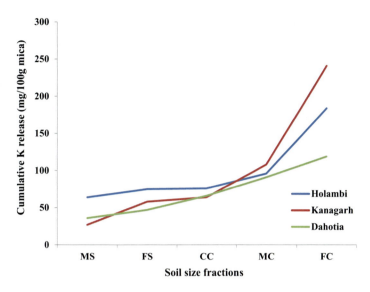

Fig. 11.1 Cumulative K release from mica of soil size fractions MS = medium silt (20-6 μm), FS = fine silt (6-2 μm), CC = coarse clay (2–0.6 μm), MC = medium clay (0.6–0.2 μm), FC = fine clay (< 0.2 μm) fractions (Adapted from Pal and Durge 1989)

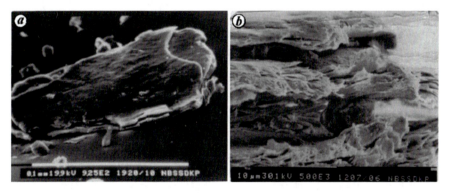

Fig. 11.2 Representative scanning electron microscopic photographs of fine-grained biotites in the semi-arid (a, b) parts of the IGP (Adapted from Pal et al. 2009b)

and particle size, a trend opposite to that of specimen minerals (Reichenbach 1972; Pal 1985), is observed and explained on the basis of edge weathering of soil mica (Pal and Durge 1989).

Vermiculite as the weathered product of biotites during the dry climates is primarily responsible for the adsorption of K in polygenetic IGP soils. However, the observed selective adsorption by smectite (Fig. 11.3) (Pal and Durge 1989), is in contrast to the behaviour of specimen mineral (Rich 1968) and soil clay smectite of dioctahedral nature (Pal and Durge 1987). Therefore, the increase in K adsorption with the increase in smectite content (Fig. 11.4) does suggest that in addition to

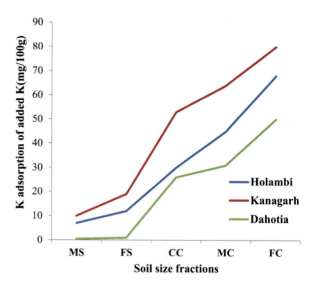

Fig. 11.3 Adsorption of added K in different soil size fractions of representative IGP soils *MS* = medium silt (20-6 μm), *FS* = fine silt (6-2 μm), *CC* = coarse clay (2–0.6 μm), *MC* = medium clay (0.6–0.2 μm), *FC* = fine clay (< 0.2 μm) fractions (Adapted from Pal and Durge 1989)

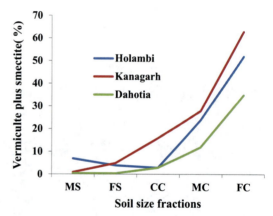

Fig. 11.4 Vermiculite plus smectite content in different soil size fractions of representative IGP soils *MS* = medium silt (20-6 μm), *FS* = fine silt (6-2 μm), *CC* = coarse clay (2–0.6 μm), *MC* = medium clay (0.6–0.2 μm), *FC* = fine clay (< 0.2 μm) fractions (Adapted from Pal and Durge 1989)

trioctahedral vermiculite, clay size trioctahedral smectites which are also the weathered product of biotites during the latter semi- arid climates also adsorbed K selectively. The trioctahedral smectites, as defined according to Alexiades and Jackson (1965), expanded to 1.7 nm peak on glycolation but contracted readily to 1.0 nm on K-saturation at 110 °C indicating it to be a high charge smectite or low charge vermiculite. Therefore, the observed K adsorption behaviour demonstrates the critical role of layer charge density in the adsorption of K by clay smectites of the many IGP soils developed in the micaceous alluvium of Himalayan origin.

11.3 K Release and Adsorption in Polygenetic Vertisols

Smectitic parent alluvium derived from the weathering Deccan basalt, deposited during a previous humid climate and in such alluvium, the cracking clay soils (Vertisols and vertic intergrades) were developed in SAT climate of the Holocene period. The fine clay smectites are of low charge di-octahedral nature, fairly well-crystallized, and do not show any sign of transformation except for low hydroxy-interlayering in the smectite interlayers (Srivastava et al. 2002; Pal et al. 2009b, 2012). The subsoils of SAT Vertisols remained under less amount of water as compared to those of wetter climates during the Holocene. Soils of SAT parts of Peninsular India were modified as a result of formation of pedogenic calcium carbonate, subsoil sodicity, poor plasma separation and cracks cutting through the slickensided zones (See Chap. 3, Figs. 3.2b, c; Chap. 4, Figs. 4.3 and 4.4). Thus they qualify as polygenetic soils (Pal et al. 2001b, 2009a, 2012). The Deccan basalt does not contain micas (Pal and Deshpande 1987). The small amounts of micas (Fig. 11.5) in Vertisols are concentrated mainly in their silt and coarse clay fractions (Pal and Durge 1987; Pal et al. 2001a) and their parental legacy is ascribed to erosional and depositional episodes experienced by the Deccan basalt areas during the post Plio-Pleistocene transition period (Pal and Deshpande 1987). The cumulative K release through the Ba-K exchange reaction, of sand, silt and clay fractions and their biotite contents provide incontrovertible evidence that K release in Vertisols is primarily controlled by biotite mica (Pal et al. 2006) even in presence of muscovite mica (Pal et al. 2001a). Cumulative amounts of K released from sand, silt and clay biotite (Fig. 11.6) indicate that, in general, clay biotite released the highest amount of K. But it is interesting to note that not only the sand-sized but also some portion of silt-sized biotite also released substantial amount of K, which are almost comparable because both sand and silt biotites have a greater number of elementary layers along with little weathered biotite (Fig. 11.5). The relevance of the almost-unweathered biotites is that both sand and silt biotites have highly favourable K release potential (Pal et al. 2006). Therefore, the released amounts of K from sand-, silt- and clay-sized biotite is in contrast to the relationships observed between

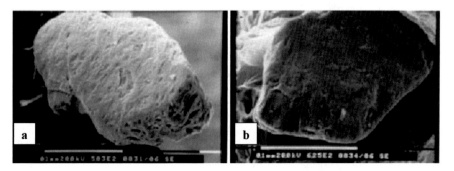

Fig. 11.5 Representative SEM photographs showing no or very little alteration of muscovite (**a**) and biotite (**b**), of Vertisols of Peninsular India (Adapted from Pal et al. 2006)

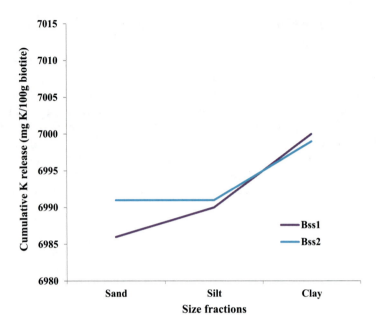

Fig. 11.6 Cumulative K release from sand (500-50 μm), silt (50–2 μm) and clay size (< 2 μm) biotites of a representative SAT Vertisols (Adapted from Pal et al. 2006)

cumulative K release and particles of specimen biotite by earlier researchers (Pal 1985; Pal et al. 2001a; Reichenbach 1972).

Potassium adsorption in Vertisols is not as severe as their clay smectites adsorbed 50–60% of added K and the amount of K adsorption by the silt and clay fraction is due to the presence of vermiculite (Fig. 11.7) and not smectite (Pal and Durge 1987). The smectites of Vertisol clays belong to the low-charge di-octahedral type and are the product of plagioclase feldspar during previous humid climate (Pal et al. 2012). Thus they have no K selectivity (Brindley 1966). This fact is evident from the almost comparable amount of K adsorption by coarse (2–.02 μm) and fine clay (0.2 μm) fractions (Fig. 11.8) even when the fine clay contains the highest amount of low-charge dioctahedral smectite (Fig. 11.9). On the other hand, silt and clay trioctahedral vermiculites are the weathered product of sand and silt- sized biotites during the post depositional period of the smectitic parent material in semi-arid Holocene period and adsorbed K from the added K (Fig. 11.7). The content of fine clay trioctahedral vermiculites was quantified following the method of Alexiades and Jackson (1965), and the vermiculite content ranged from 5 to 9% (Pal and Durge 1987) (Fig. 11.7). Because of their low content vermiculites are generally not detected on the glycolation of Ca- saturated samples but can be detected by a progressive reinforcement of the 1.0 nm peak of mica while heating the K saturated samples from 25 to 550 °C (Pal and Durge 1987).

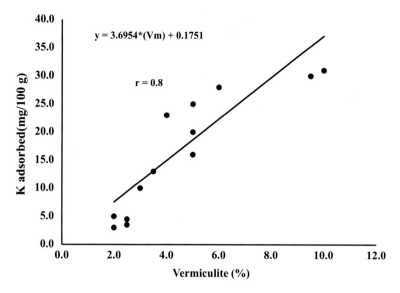

Fig. 11.7 Relation between K adsorbed and vermiculite content of soil size fractions of representative SAT Vertisols (Adapted from Pal and Durge 1987)

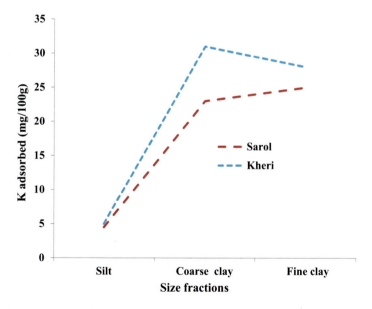

Fig. 11.8 Adsorption of added K in silt (50-2 μm), coarse clay (2–0.2 μm), fine clay (<0.2 μm) fractions of representative SAT Vertisols (Adapted from Pal and Durge 1987)

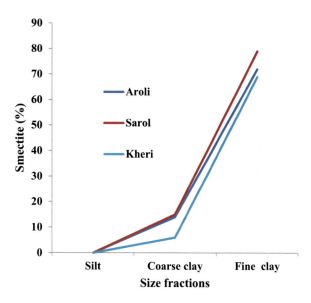

Fig. 11.9 Smectite content in the silt (50-2 µm), coarse clay (2–0.2 µm), fine clay (< 0.2 µm) fractions of representative SAT Vertisols (Adapted from Pal and Durge 1987)

11.4 K Release and Adsorption in Polygenetic RF Soils

It was so long believed that the predominance of kaolinite followed by illite maintains the low nutrient holding capacity of RF soils (Alfisols) of the southern Peninsular India; and therefore application of balanced fertilizer including K is generally recommended. However, in many of these soils, crops do not respond to K fertilizer application (Ghosh and Biswas 1978; Rego et al. 1986) because many SAT Alfisols contain sufficient biotite mica (Pal et al. 2014).

Recent paleopedological studies demonstrated that these RF Alfisols of the SAT are relict paleosols of polygenetic nature. The upper layers of the soils of the preceding tropical humid climate were truncated by multiple arid erosional cycles due to climate change from tropical humid to semi-arid during the Plio-Pleistocene transition period (See Chap. 5, Fig.5.1). Erosion exposed the relatively less weathered subsoils wherein considerable amount of unaltered biotite particles remained in the sand and silt fractions (Fig. 11.10) (Pal and Durge 1989; Pal et al. 1993). Therefore, many of the proposed relationships between K release and mica particle size hitherto obtained either in weathered soils or specimen minerals may not be valid in these polygenetic soils (Pal et al. 1993).

The cumulative K release from different size fractions of two benchmark RF soils (Alfisols) of SAT namely Patancheru and Nalgonda of southern India, indicated a contrasting particle size-K release relationship between silts and clays, since K release increased with the fineness of clay size biotite mica but decreased with the fineness of silt size mica (Fig. 11.11). The zones in clay biotites are at different stages of expansion, and under such circumstances K release would occur mainly by edge weathering as is evident from increased K release with the decrease in mica

11.4 K Release and Adsorption in Polygenetic RF Soils

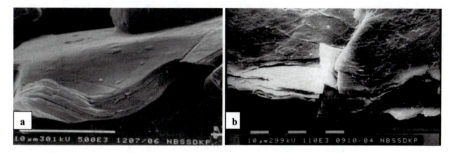

Fig. 11.10 Representative SEM photographs of almost unaltered biotites with minor layer separation and bending of edges in SAT Alfisols (**a**), Patancheru, Andhra Pradesh and (**b**), Dyavapatna, Karnataka (Adapted from Pal et al. 2000)

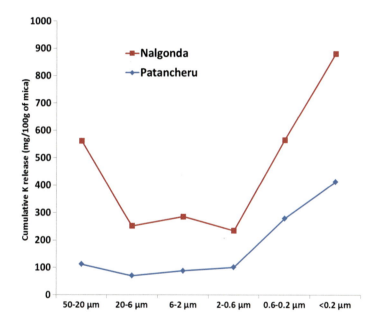

Fig. 11.11 Potassium release from various size fractions of RF Alfisols (Patancheru and Nalgonda) of southern India (Adapted from Pal et al. 1993)
CS = coarse silt (50-20 µm), *MS* = medium silt (20-6 µm), *FS* = fine silt (6-2 µm), *CC* = coarse clay (2–0.6 µm), *MC* = medium clay (0.6–0.2 µm), *FC* = fine clay (<0.2 µm)

particles (Pal and Durge 1989). On the other hand the K release increased with the increase in silt sized mica because of the almost unweathered sand and silt biotites (Pal et al. 1993, 2014); this particular K release trend is normally obtained with specimen micas (Pal 1985). Thus, quite favourable K release rate from both silt and clay biotites explains as to why crop response to fertilizer K is seldom obtained in many of RF Alfisols under SAT environments (Pal et al. 2014).

Although enriched with biotite mica, many SAT Alfisols are not endowed sufficiently with trioctahedral vermiculite but enriched with smectite (Fig. 11.12).

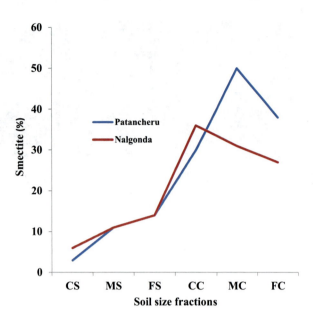

Fig. 11.12 Smectite content of soil size fractions of representative RF Alfisols of SAT regions CS = coarse silt (50–20 µm), MS = medium silt (20-6 µm), FS = fine silt (6-2 µm), CC = coarse clay (2–0.6 µm), MC = medium clay (0.6–0.2 µm), FC = fine clay (< 0.2 µm) fractions (Adapted from Pal et al. 1993)

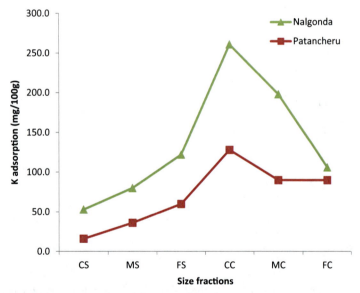

Fig. 11.13 Potassium adsorption by various size fractions of RF Alfisols of the SAT regions (Patancheru and Nalgonda, Andhra Pradesh, India), *CS* = coarse silt (50-20 µm), *MS* = medium silt (20-6 µm), *FS* = fine silt (6-2 µm), *CC* = coarse clay (2–0.6 µm), *MC* = medium clay (0.6–0.2 µm), *FC* = fine clay (<0.2 µm) (Adapted from Pal et al. 1993; Pal and Durge 1989)

Therefore, the observed K adsorption by silt and clay fractions of benchmark RF Alfisols of Patancheru and Nalgonda series (Fig. 11.13) may be attributed to smectite content because it increases with the decrease in soil size fractions (Fig. 11.12) (Pal et al. 1993). However, perusal of the K adsorption data (Fig. 11.13) indicates that despite sufficient amount of smectite, the finer fractions of clay particularly, the medium (0.6–0.2 µm) and fine clay fractions (< 0.2 µm) does not adsorb K proportionately. This suggests that clays contain both high and low charge smectite and a considerable proportion of smectite is a low charge dioctahedral smectite, which does not have K selectivity (Pal et al. 1993). The co-existence of both low and high charge smectite is however related to their respective genesis in paleoclimatic environments (Pal et al. 1989).

Pal et al. (1989) demonstrated that a part of the well crystallized clay size dioctahedral smectite was the first weathering product of granite-gneiss that survived transformation to kaolin in a Pre-Pliocene tropical humid climate and is preserved to the present along with kaolin. The biotite which somehow survived HT climate weathering, altered to trioctahedral smectite mainly in the silt and partially in the coarse clay fractions under the semi-arid climate of the Plio-Pleistocene transition period. The results on the selective K adsorption by trioctahedral smectite and not by dioctahedral smectite are in accord with those reported earlier with high and low charge smectites respectively (Pal and Durge 1987, 1989). SAT Alfisols are generally rich in clay and therefore, it is envisaged that these soils may have less K fixation problem (Pal et al. 2000).

The above discussions suggest that the observed anomalous K release and adsorption reactions appear to be unique when these reactions are explained in terms of the genesis and transformation of 2:1 layer silicate minerals during the Holocene climate change. Thus, these unique reactions provide valuable information about the polygenesis and also guidelines in K managements of Indian tropical soils. Methods used to follow the K release and adsorption reactions are simple as described by Pal and Durge (1987, 1989) and Pal et al. (1993).

References

Alexiades CA, Jackson ML (1965) Quantitative determination of vermiculite in soils. Soil Sci Soc Amer Proc 29:522–527
Brindley GW (1966) Ethylene glycol and glycerol complexes of smectites and vermiculites. Clay Miner 6:237–259
Ghosh AB, Biswas CR (1978) Potassium responses and changes in soil potassium status with time. Potassium in soils and crops. Potash Research Institute of India, Delhi, pp 379–390
Jenkins DA (1985) Chemical and mineralogical composition in the identification of paleosols. In: Boardman J (ed) Soils and quaternary landscape evolution. Wiley, New York, pp 23–43
Pal DK (1985) Potassium release from muscovite and biotite under alkaline conditions. Pedologie (Ghent) 35:133–146
Pal DK, Deshpande SB (1987) Characteristics and genesis of minerals in some benchmark Vertisols of India. Pedologie (Ghent) 37:259–275
Pal DK, Durge SL (1987) Potassium release and fixation reactions in some benchmark Vertisols of India in relation to their mineralogy. Pedologie (Ghent) 37:103–116

Pal DK, Durge SL (1989) Release and adsorption of potassium in some benchmark alluvial soils of India in relation to their mineralogy. Pedologie (Ghent) 37:103–116

Pal DK, Deshpande SB, Venugopal KR, Kalbande AR (1989) Formation of di and trioctahedral smectite as an evidence for paleoclimatic changes in southern and central peninsular India. Geoderma 45:175–184

Pal DK, Deshpande SB, Durge SL (1993) Potassium release and adsorption reactions in two ferruginous soils(polygenetic)soils of southern India in relation to their mineralogy. Pedologie (Ghent) 43:403–415

Pal DK, Bhattacharyya T, Deshpande SB, Sarma VAK, Velayutham M (2000) Significance of minerals in soil environment of India. NBSS review series, 1. NBSS&LUP, Nagpur, p 68

Pal DK, Srivastava P, Durge SL, Bhattacharyya T (2001a) Role of weathering of fine grained micas in potassium management of Indian soils. Applied Clay Sci 20:39–52

Pal DK, Balpande SS, Srivastava P (2001b) Polygenetic Vertisols of the Purna Valley of Central India. Catena 43:231–249

Pal DK, Nimkar AM, Ray SK, Bhattacharyya T, Chandran P (2006) Characterisation and quantification of micas and smectites in potassium management of shrink–swell soils in Deccan basalt area. In: Benbi DK, Brar MS, Bansal SK (eds) Balanced fertilization for sustaining crop productivity. Proceedings of the International Symposium held at PAU, Ludhiana, India, 22–25 Nov' 2006 IPI, Switzerland, pp 81–93

Pal DK, Bhattacharyya T, Chandran P, Ray SK, Satyavathi PLA, Durge SL, Raja P, Maurya UK (2009a) Vertisols (cracking clay soils) in a climosequence of peninsular India: evidence for Holocene climate changes. Quatern Int 209:6–21

Pal DK, Bhattacharyya T, Srivastava P, Chandran P, Ray SK (2009b) Soils of the indo-Gangetic Plains: their historical perspective and management. Curr Sci 9:1193–1201

Pal DK, Wani SP, Sahrawat KL (2012) Vertisols of tropical Indian environments: pedology and edaphology. Geoderma 189–190:28–49

Pal DK, Wani SP, Sahrawat KL, Srivastava P (2014) Red ferruginous soils of tropical Indian environments: a review of the pedogenic processes and its implications for edaphology. Catena 121:260–278. https://doi.org/10.1016/j.catena2014.05.023

Rego TJ, Sahrawat KL, Burford JR (1986) Depletion of soil potassium in an Alfisol under improved rainfed and cereal/legume cropping system in the Indian SAT. Trans 13th Int Congr Soil Sci 3:928–929

Reichenbach HGV (1972) Factors of mica transformation. Proceedings 9th Colloquium International Potash Institute, pp 33–42

Rich CI (1968) Mineralogy of soil potassium. In: Kilmer et al (eds) The role of potassium in agriculture. American Society of Agronomy, Madison, pp 79–108

Srivastava P, Parkash B, Pal DK (1998) Clay minerals in soils as evidence of Holocene climatic change, central indo-Gangetic Plains, north-central India. Quat Res 50:230–239

Srivastava P, Bhattacharyya T, Pal DK (2002) Significance of the formation of calcium carbonate minerals in the pedogenesis and management of cracking clay soils (Vertisols) of India. Clay Clay Miner 50:111–126

Srivastava P, Pal DK, Aruche KM, Wani SP, Sahrawat KL (2015) Soils of the indo-Gangetic Plains: a pedogenic response to landscape stability, climatic variability and anthropogenic activity during the Holocene. Earth-Sci Rev 140:54–71. https://doi.org/10.1016/j.earscirev.2014.10.010

Srivastava P, Aruche M, Arya A, Pal DK, Singh LP (2016) A micromorphological record of contemporary and relict pedogenic processes in soils of the Indo-Gangetic Plains: implications for mineral weathering, provenance and climatic changes. Earth Surf Proc Land 41:771–790. https://doi.org/10.1002/esp.3862

Chapter 12
Concluding Remarks

Abstract The book 'Simple methods to study pedology and edaphology of tropical Indian soils' closes with a chapter 'Concluding Remarks'. It has been always difficult to manage tropical soils to sustain their productivity because comprehensive knowledge on their formation remained elusive for a long time. Soil care has become a national agenda in the Indian context to meet the food demand for ever increasing human population. This task demands the basic pedological research to resolve some of the enigmatic edaphological aspects of soils to develop improved management practices. During the last few decades Indian pedologists provided insights into pedology, paleopedology, polygenesis, mineralogy, taxonomy and edaphology when the focus of soil research changed qualitatively due to the use of geomorphic and climatic history of landscapes alongside the use of high resolution mineralogical, micromorphological and age-control tools. Expansion of basic and fundamental knowledge base on Indian tropical soils provided unique guiding principles to develop several index soil properties as simple diagnostic analytical methods to study the pedology and edaphology of major soil types of India. This chapter highlights the major theme areas of soils (Chaps. 2, 3, 4, 5, 6, 7, 8, 9, 10, 11) that have been dealt in the perspective of the recent developments in simple methods to study pedology and edaphology.

Keywords Indian tropical soils · Pedology · Edaphology · Analytical methods

For the past several years, many thought that the tropical soils such as the deep red and highly weathered soils are either agriculturally poor or virtually useless. In India, Vertisols, Mollisols, Alfisols, Ultisols, Aridisols, Inceptisols and Entisols exist in more than one bio-climatic zones of India, and the absence of Oxisols, suggests that soil diversity in India is large and therefore, any generalizations about tropical soils are unlikely to have wider applicability in the Indian subcontinent. These soils are not confined to a single production system and contribute substantially to India's growing self-sufficiency in food production and food stocks.

Although much valuable work has been done throughout the tropics, it has been always difficult to manage these soils to sustain their productivity, and it is more so when comprehensive knowledge on their formation remained incomplete for a long time. Therefore, soil care needs to be a constant research endeavour in Indian tropical environment as new soil knowledge base becomes critical when attempts are made to fill the gap between food production and future population growth. In this task basic pedological research is needed to understand some of the unresolved edaphological aspects of the tropical Indian soils to develop improved management practices. Realizing this urgency, research endeavours during the last few decades on benchmark and identified soil series by the Indian pedologists and earth scientists were made to provide insights into several aspects of five pedogenetically important soil orders like Alfisols, Mollisols, Ultisols, Vertisols and Inceptisols of tropical Indian environments, with a special attention on the pedology, paleopedology, polygenesis, mineralogy, taxonomy and edaphology. A much better knowledge was made available when the focus of soil research changed qualitatively due to the use of geomorphic history of landscapes alongside the use of high resolution mineralogical, micromorphological and age-control tools. Such advancement of knowledge base helped to measure the relatively subtle processes related to pedology of the present and past geological periods, and also on mineralogical transformations and their impact in rationalizing the soil taxonomic database. Expansion of basic and fundamental knowledge base on Indian tropical soils provided unique guiding principles to develop several index soil properties as simple diagnostic methods to study the pedology and edaphology of major soil types of India.

India being mainly as an agrarian society, soil care through recommended practices of the National Agricultural Systems (NARS) has been the engine of economic development. However, to facilitate the economic development, application of recommended practices of the National Agricultural Research Systems (NARS) is required to be continuous and sustaining. To fulfil this national task, adequate basic pedological database is to be developed as soon as possible. To fulfil the present gap in such information, a fresh research initiative is warranted though it is not possible without the help of sophisticated instrumental facilities to study pedology and edaphology. To circumvent this problem a need was felt to develop simple methods based on the guiding principles emerged from rigorous basic pedological research that were undertaken at many national research institutes for the last few decades.

In view of the above background, a total of 10 chapters were chosen that deal with very important themes of pedology and edaphology. In each chapter, a specific theme is highlighted to showcase the worthiness of applying such simple analytical method that has emerged from rigorous research endeavours for the last several decades mainly at the ICAR-NBSS&LUP, Nagpur, India. It is hoped that such simple methods as described in this communication will help to unravel many pedological, edaphological, mineralogical and taxonomical issues of Indian tropical soils, and to develop the most sought after the robust soil information system for better management and optimization of soil productivity for populous Indian subcontinent and other countries in the developing world in the twenty-first century.

Index

A
Acidity, 49, 51, 52, 67–70, 77, 78
Alfisols, vii, 2, 3, 10, 12–15, 23, 33–36, 43, 44, 49–51, 54, 64–69, 75–78, 82, 88–91, 93, 94
Analytical methods, 3, 9–10, 43–44, 64, 67, 69, 78, 94
Argillic horizon, 8, 10–16, 23, 33
Argilli-turbation, 20

B
Biotite, 8, 22, 36, 70–71, 82, 83, 85, 86, 88, 89, 91

C
Calcium carbonate, 28, 36, 39, 41–42, 59, 61, 85
Carbon, 3, 11, 20, 52, 59, 64–68
Characterisation of clay minerals, 63, 66
Clay illuviation, 3, 8–16, 19–24, 33, 35
Climate change, vii, 3, 27–29, 36, 82, 88, 91
Climosequence, 28, 29
Cracking depths, 27–29

D
Deep cracking clay soils, 42, 54, 61, 70, 71, 85
Drainage, 41, 44, 45, 59, 61

E
Edaphology, vii, viii, 1–3, 42, 94
Erosion by water, 75–77
Evaluation for deep rooted crops, 42, 59–61

G
Genesis of hydroxy-interlayered clay minerals, 63, 66, 67
Gibbsite, 50, 51, 68, 69, 77, 78
Gypsum, 39–41, 43, 59

H
High and low charge smectite, 84, 91
High and low vermiculite, 70, 71, 82–84, 86
Holocene climate change, 3, 27–30, 91
Hydroxy-interlayered minerals, 50, 51, 65–70, 77

I
Inceptisols, vii, 2, 3, 8, 50, 54, 75, 77, 78, 93
Indian tropical soils, vii, 1–3, 14–16, 49–51, 54, 64–68, 70–71, 78, 91, 94
Indo-Gangetic Plain (IGP) soils, 8–16, 40, 41

K
Kaolin (Kl), 14, 22, 35, 36, 50, 51, 66, 76–78, 91

M
Mica, 10, 21, 50, 51, 70–71
Mineralogy class, 49, 51, 54, 67, 78
Mineralogy class of soils, 49, 51, 54, 67, 78
Mollisols, vii, 2, 3, 10, 12–14, 43, 64–67, 75–78, 93, 94

N
Nitrogen, 70–71
Nutrient management practices, 39, 64, 78, 94

P
Paleopedology, 88, 94
Palygorskite, 39, 44–46, 59
Pedogenesis, 8, 13, 14, 27, 28, 35, 51, 78
Pedogenic calcium carbonate (PC), 21–24, 28, 29, 36, 39–42, 59, 61, 77, 85
Pedology, vii, viii, 2, 3, 94
Pedoturbation, 3, 19–24
Phosphorus (P), 69
Polygenesis, 3, 82–91, 94
Polygenetic Indian tropical soils, 82–85, 88
Potassium, 70–71, 82–91
    adsorption, 70–71, 82, 85, 86, 88, 90, 91
    release, 82–91

R
Red ferruginous (RF) soils, 8–16, 41, 49–50, 69, 70, 77, 82
Regressive pedogenesis, 27, 28

S
Saturated hydraulic conductivity (sHC), 22, 28, 41, 61
SAT Vertisols, 19–24, 36, 40–42, 44, 46, 54, 57–62, 64, 66, 69, 85–88
Semi-arid (SA) climate, 2, 10–11, 21, 27, 28, 36, 64, 82, 83, 91
Semi-arid (SAT) and Humid (HT) tropical climates, 2, 10–11, 21, 27, 28, 41, 64, 82
Simple methods, vii, viii, 3, 46, 94
Simple methods for identification, 46

Smectite(Sm), 9, 10, 21, 22, 29, 35, 36, 45, 50, 51, 54, 65–70, 77, 82–86, 88, 90, 91
Soil degradation, 3, 40, 75–79
Soil management, 2, 39, 40, 42–46, 59, 64, 70, 71, 78, 94
Soil minerals, 51, 63, 69, 88
Soil modifiers, 39–46, 59, 66
Soils with clay-enriched B-horizon, 8
Southern India, 12, 36, 40, 50, 67, 88, 89
Study of pedology and edaphology, vii, viii, 2, 3, 94

T
Taxonomy, vii, 3, 10, 11, 14, 23, 24, 29, 51, 94
Tropical Vertisols, 28–30, 40–43, 65, 94
Truncation of soil profile, 34, 36

U
Ultisols, vii, 2, 3, 10–14, 49–51, 64–70, 75–78, 93, 94
Use of simple methods, vii, viii, 3, 94

V
Vertisols, vii, 2, 3, 19–24, 27–30, 35, 36, 40–46, 54, 57–62, 64–70, 82, 85–88, 93, 94

W
Weathering, 8, 13, 21, 22, 28, 34–36, 42, 51, 67, 69, 70, 77, 82, 83, 85, 88, 91

Z
Zeolites, 39, 42–44, 59, 66, 68, 69